# Eclipse Almanac
# 2041 To 2050

## Ten Years of Solar and Lunar Eclipses

### BLACK AND WHITE EDITION

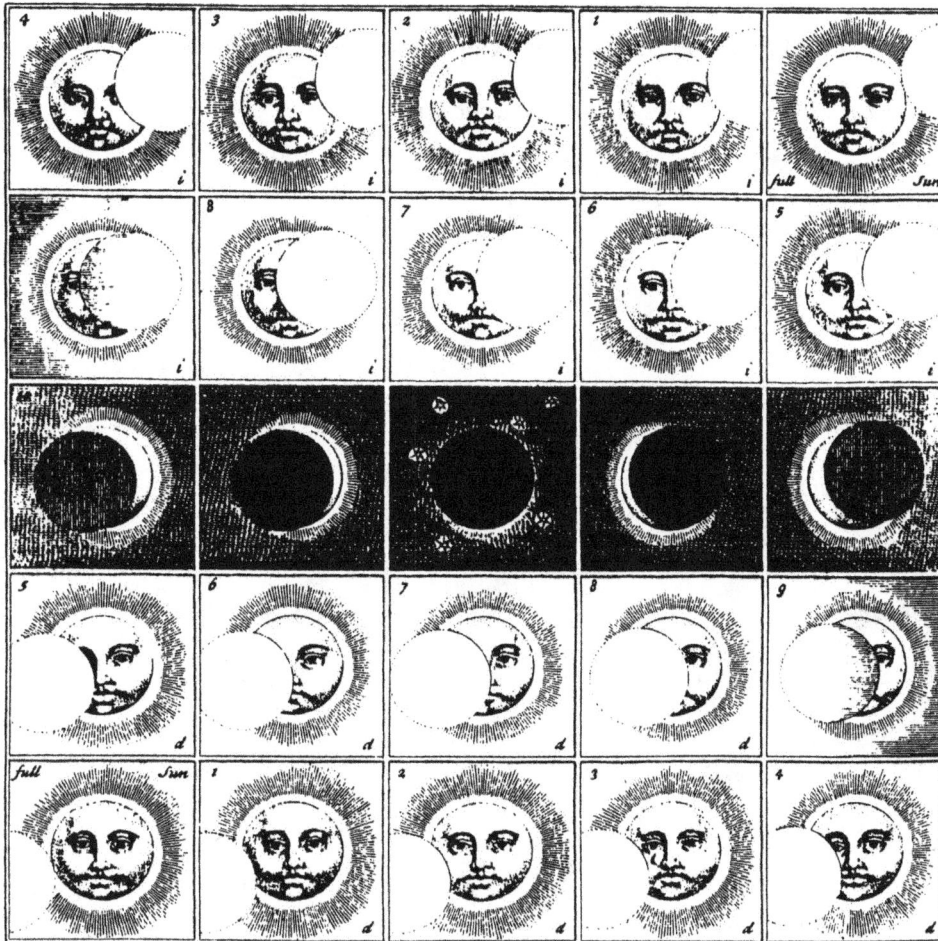

## Fred Espenak

Edition 1.0
October 2020

Eclipse Almanac 2041 To 2050 – Black and White Edition

Ten Years of Solar and Lunar Eclipses

Astropixels Publishing
P.O. Box 16197
Portal, AZ 85632

*www.astropixels.com/pubs*

This book may be ordered at: *www.astropixels.com/pubs/EclipseAlmanac.html*

More about solar and lunar eclipses from 2041 To 2050 can be found at EclipseWise:

*www.astropixels.com/news/EclipseAlmanac.html*

Astropixels Publication Number:  AP029

First Edition (Version 1.0a)

ISBN  978-1-941983-29-4

Printed in the United States of America

Front Cover: Broadside by "H. M." entitled "A plain Description of the Sun's appearance, in the Increase and Decrease of the Eclipse, which will happen on Fryday (in the morning) April the 22d, 1715."

Back Cover: Portrait of Fred Espenak (Copyright ©2018 by Fred Espenak).

# Preface

The ***Eclipse Almanac*** contains maps and diagrams of every solar and lunar eclipse over a ten-year period. This permits the reader to look ahead and easily determine when and where each of these events will be seen. Particular details about each eclipse are included as well as a 25-year table looking further into the future.

Section 1 covers solar eclipses, while Section 2 is devoted to lunar eclipses. Brief explanations are given for the different types of eclipses, and descriptions of the visual appearance of each one is included. Section 3 tabulates the date and time of the Moon's phases over a decade. New Moon and Full Moon phases that coincide with solar and lunar eclipses, respectively, are identified.

The ***Eclipse Almanac*** series consists of five volumes. Each one covering a single decade as follows:

       1) Eclipse Almanac 2021 to 2030
       2) Eclipse Almanac 2031 to 2040
       3) Eclipse Almanac 2041 to 2050
       4) Eclipse Almanac 2051 to 2060
       5) Eclipse Almanac 2061 to 2070

Each volume is available in a color edition as well as a more economical black and white edition.

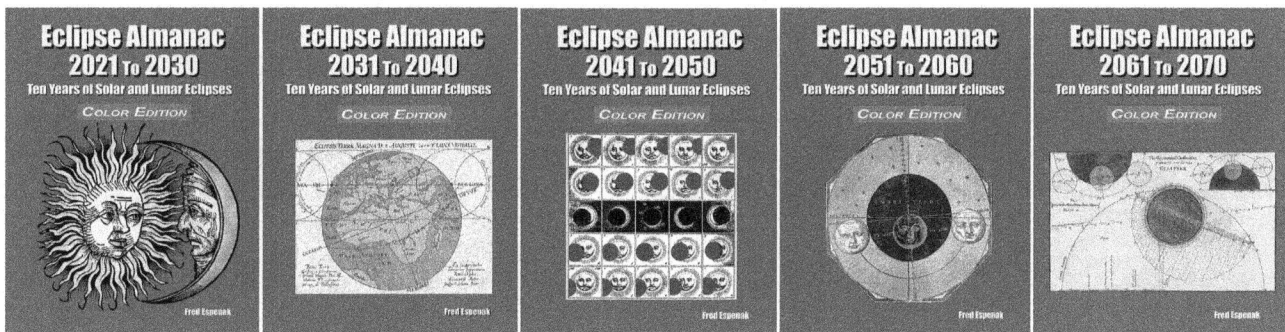

All times listed in the ***Eclipse Almanac*** are in Universal Time (UT1). This is the modern replacement of Greenwich Mean Time, which was based on mean solar time from Greenwich, England. In comparison, Universal Time is based on Earth's rotation with respect to distant quasars.

For North Americans, the conversion from UT1 to local time is as follows:

       Atlantic Standard Time (AST)    = UT1 - 4 hours
       Eastern Standard Time (EST)    = UT1 - 5 hours
       Central Standard Time (CST)    = UT1 - 6 hours
       Mountain Standard Time (MST)  = UT1 - 7 hours
       Pacific Standard Time (PST)    = UT1 - 8 hours

If Daylight Saving Time is in effect in the time zone, you must ADD one hour to the standard time.

The conversion of Universal Time to other time zones is easily found on the Internet.

---

For more information about solar and lunar eclipses from 2021 to 2070, visit *EclipseWise.com* at:

    *www.astropixels.com/news/EclipseAlmanac.html*

## Table of Contents

## Central Solar Eclipses from 2041 to 2050

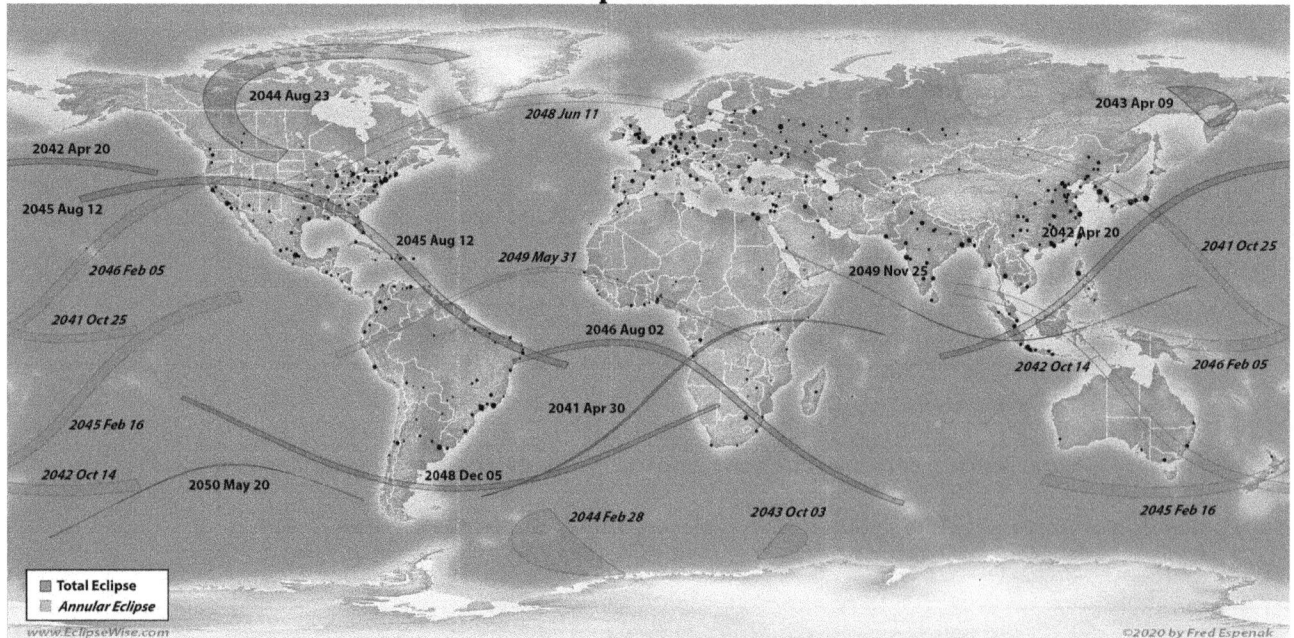

The path of every central solar eclipse from 2041 to 2050 is plotted on a world map. The central paths of total eclipses are shaded blue, while annular eclipses are shaded red. For hybrid eclipses, part of the path is shaded blue (total), and part is shaded red (annular). Major cities are plotted as black dots, scaled by population size. (©2020 F. Espenak)

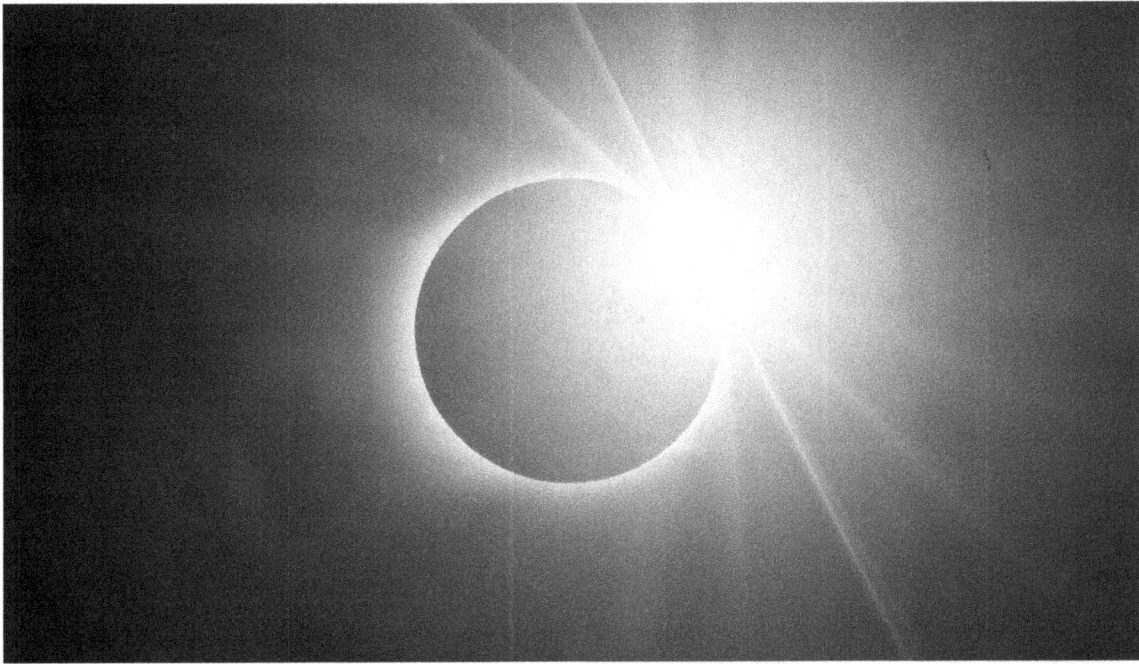

*Photo 1–1 The "Diamond Ring Effect" is seen just before totality begins. Total Solar Eclipse of 2018 Jul 03. ©2018 F. Espenak*

# Section 1: Solar Eclipses

## Introduction

The Moon orbits Earth once every 29.5306 days with respect to the Sun. Over the course of its orbit, the Moon's changing position relative to the Sun results in its familiar phases: New Moon > First Quarter > Full Moon > Last Quarter > New Moon. The New Moon phase is the only one not visible because the illuminated side of the Moon then points away from Earth.

The orbit of the Moon is tilted about 5.1° to Earth's orbit around the Sun. The points where the two orbits appear to cross are called the nodes. When the New Moon occurs near one of these nodes, the Moon's shadow falls on a portion of Earth and a solar eclipse is visible from that region.

The Moon's shadow is composed of three cone-shaped components. The outer or penumbral shadow is a zone where the Sun's rays are partially blocked. Nested within the penumbra is the umbral shadow — a region where direct rays from the Sun are completely blocked. The conical umbra tapers to a point beyond which extends an expanding cone called the antumbra. From within this third shadow, the Moon appears smaller than the Sun and is seen in silhouette against the solar disk.

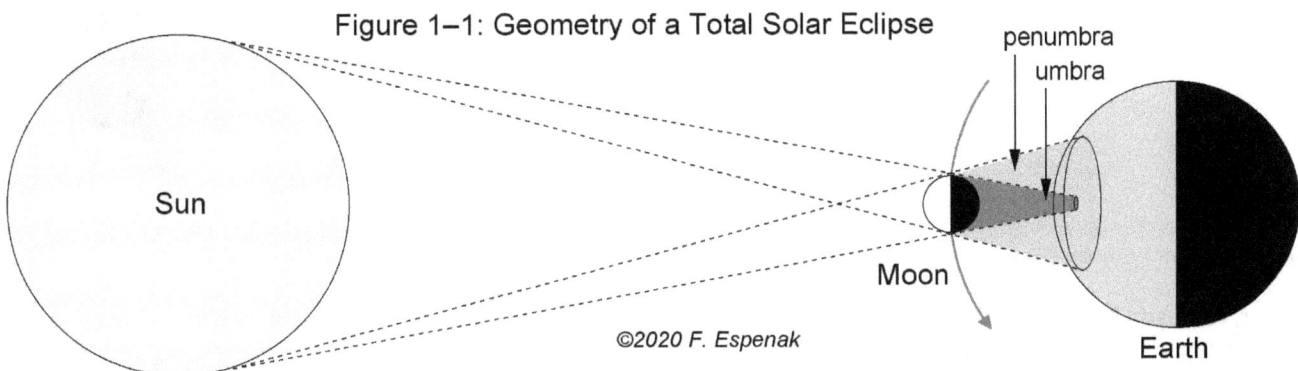

*Figure 1–1 illustrates the geometry of a total solar eclipse. A partial eclipse is visible from within the large penumbral shadow, while the total eclipse is only seen from the much smaller umbral shadow.*

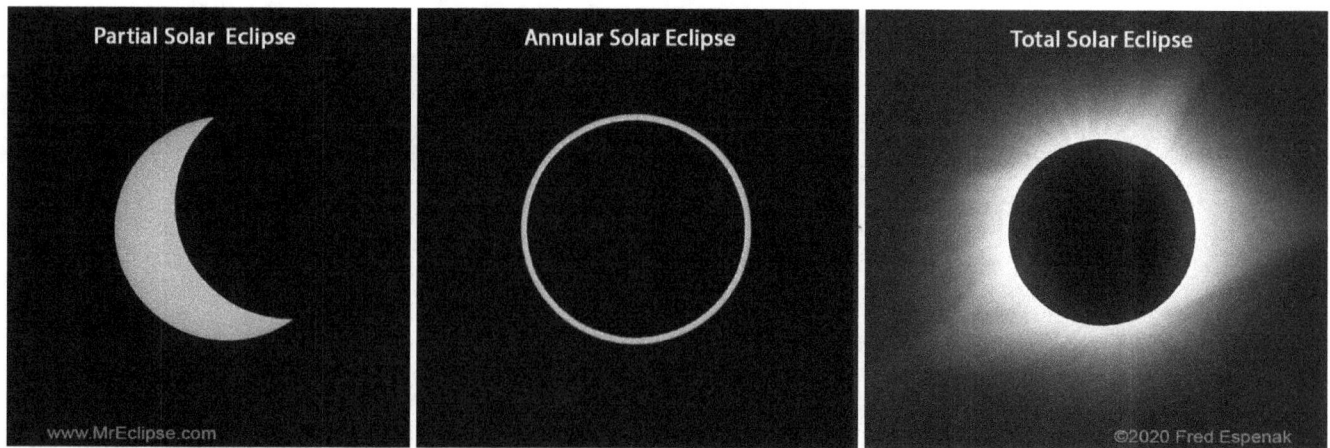

*Photo 1–2 Examples of a partial, annular, and total solar eclipse. ©2020 F. Espenak*

## Types of Solar Eclipses

There are four types of solar eclipses:

1. **Total Solar Eclipse** — The Moon's penumbral and umbral shadows traverse Earth. The Moon's antumbral shadow extends beyond Earth's surface. The Moon appears larger than the Sun and completely covers the solar disk when viewed from within the umbral shadow. A partial eclipse is seen within the penumbral shadow (Figure 1–1).
2. **Annular Solar Eclipse** — The Moon's penumbral and antumbral shadows traverse Earth. The Moon's umbral shadow completely misses Earth. The Moon's disk appears smaller than the Sun so a bright ring of the Sun's disk surrounds the Moon when viewed from within the antumbral shadow. A partial eclipse is seen within the penumbral shadow (Figure 1–2).
4. **Hybrid Solar Eclipse** — The Moon's penumbral, umbral and antumbral shadows all traverse parts of Earth. The curvature of Earth's surface brings some regions into the umbra and others into the antumbra. The eclipse is total within the umbra and annular within the antumbra. A partial eclipse is seen within the penumbral shadow. Hybrid eclipses are also known as annular-total eclipses.
4. **Partial Solar Eclipse** — The Moon's penumbral shadow traverses Earth while the umbral and antumbral shadows miss Earth. A portion of the Sun's disk is obscured from within the penumbra.

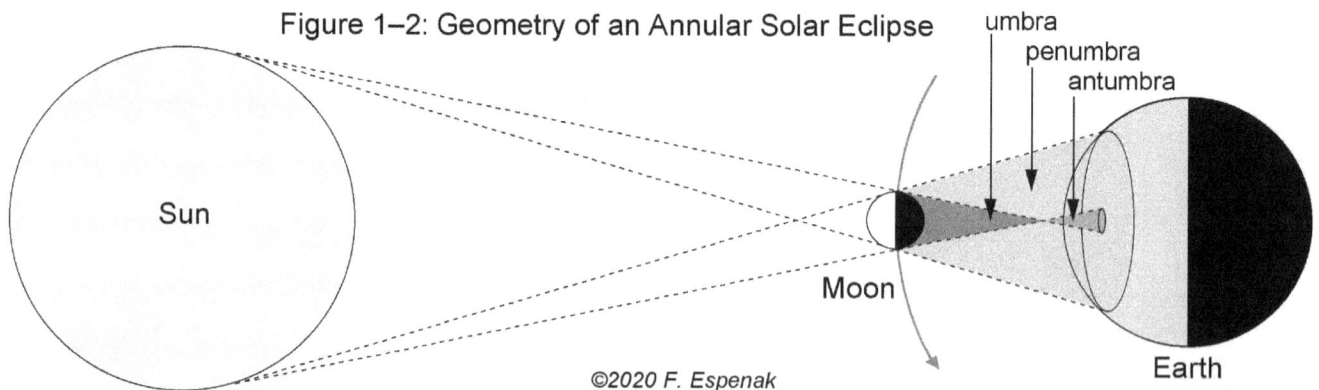

*Figure 1–2 illustrates the geometry of an annular solar eclipse. A partial eclipse is visible from within the large penumbral shadow, while the annular eclipse is confined to the much smaller antumbral shadow.*

Annular, total and hybrid eclipses are sometimes referred to as *central* eclipses[1]. However, on rare occasions it is possible to have an annular or total eclipse that is non-central. This occurs in Earth's polar regions when only the edge of the umbral or antumbral falls on Earth while the shadow axis misses the planet entirely. The eclipse of 2043 April 09 is a non-central total solar eclipse.

---

[1] A central eclipse is one in which the central axis of the Moon's umbral/antumbral shadow intersects with Earth's surface.

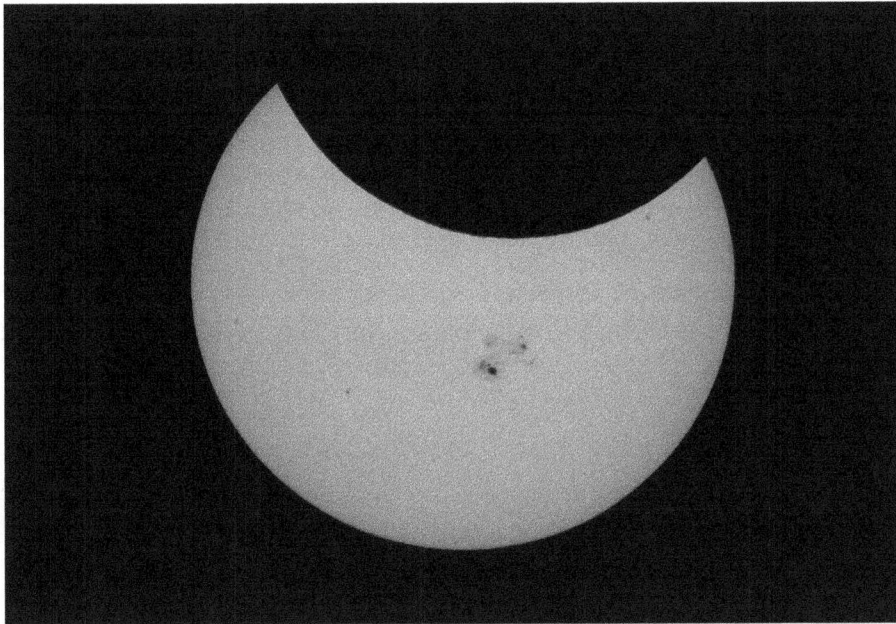

*Photo 1–3 Partial solar eclipse of 2014 Oct 23. ©2014 F. Espenak*

## Visual Appearance of Partial Solar Eclipses

When the Moon's penumbral shadow strikes Earth, a partial eclipse of the Sun is visible from that region. The apparent motion of the Moon with respect to the Sun is gradual — the partial phases can two hours or more. During this time, the Moon's dark limb slowly creeps across the Sun's disk.

Partial eclipses cannot be viewed with the unprotected eye because the Sun is still extremely bright. Special techniques are needed to safely view the eclipse (for example, special eclipse filters or glasses). Even when a partial eclipse reaches its maximum phase, the sky and landscape remain bright. But careful inspection of dappled sunlight beneath a leafy tree will reveal multiple images of the eclipse. The gaps between the tree leaves act like pinhole cameras and project images of the crescent Sun onto the ground.

The Moon's penumbral shadow is typically 4200 to 4500 miles (6700 to 7300 kilometers) in diameter and can cover a significant fraction of the daytime hemisphere of Earth. Consequently, partial eclipses are visible from large geographic regions as the penumbra sweeps across Earth's surface.

*Photo 1–4 shows various phases of the Annular solar eclipse of 2005 Oct 03. ©2005 F. Espenak*

## Visual Appearance of Annular Solar Eclipses

During an annular eclipse, the Moon's penumbral and antumbral shadows sweep across Earth. Compared to the penumbra, the antumbra is much smaller and has a maximum diameter of 232 miles (374 kilometers). Because of this, the antumbra covers a tiny fraction of Earth's surface.

A partial eclipse is visible within the penumbral shadow, but only observers located in the much narrower track of the antumbra will see an annular eclipse. For this reason, the antumbra's trajectory across Earth is called the path of annularity.

All annular eclipses begin with a series of partial phases lasting an hour or more. At the peak of the eclipse, the Moon's disk can be seen in complete silhouette against the Sun. The remaining solar photosphere appears as an intensely bright ring of light surrounding the Moon. The annular phase can last a maximum of 12 ½ minutes but is more typically 3 to 5 minutes in length. After annularity, another series of partial phases occur as the Moon gradually uncovers the Sun.

Special precautions must be used to watch an annular eclipse (just like partial eclipses). Even during the annular phase, the Sun is dangerously bright and cannot be viewed without a solar filter. The landscape and sky remain bright throughout the eclipse, giving little indication of the celestial event in progress.

*Photo 1–5 The Sun's corona is only visible to the naked eye for a few minutes during a total eclipse of the Sun.*
*This is an HDR composite image of the total solar eclipse of August 21, 2017 from Casper, Wyoming.*
*© 2017 F. Espenak, www.MrEclipse.com*

## Visual Appearance of Total Solar Eclipses

During a total eclipse, the Moon's penumbral and umbral shadows fall upon Earth. The umbra has a maximum diameter of 170 miles (273 kilometers). The narrow track traced out as the umbra sweeps across Earth's surface is called the path of totality. The Sun is completely obscured by the Moon from within this zone. The total phase can last up to 7 ½ minutes but is more typically 2 to 3 minutes in length.

Total eclipses all begin and end with a series of partial phases lasting an hour or more. But this is where the resemblance to partial and annular eclipses ends — the total phase is the most spectacular astronomical event visible to the naked eye. During totality, the Sun's outer atmosphere — the solar corona — appears as a gossamer halo surrounding the Moon, and bright stars and planets are visible.

The eclipse takes on a unique character about five minutes before the total phase commences. The Sun has a foreboding quality and casts abnormally sharp shadows. The approaching lunar umbra darkens the western sky and the air temperature is noticeably cooler. A minute before totality, pale shadow bands[2] ripple across the ground.

The ambient light grows feeble even though the crescent Sun is still too bright to see. In the final seconds, the Sun's corona emerges from the glare as the solar crescent shrinks to a brilliant jewel. This celestial diamond ring lingers for a moment before the sunlight is extinguished and totality begins.

The Sun's glorious corona is now displayed to full advantage in the darkened sky. Standing within the Moon's umbra affords the rare and unprecedented opportunity to gaze directly at the glowing million-degree plasma surrounding our star. Twisted, tortured, and constrained by the Sun's enormous magnetic fields, the solar corona is revealed to the naked eye only during the brief seconds when the Moon completely blocks the brilliant disk of the Sun. An eerie twilight bathes the landscape and the colors of dusk surround the horizon.

Minutes race by like seconds. Suddenly, a sparkling bead of sunlight reappears along one edge of the Moon and quickly grows to blindingly bright proportions. Daylight returns as the corona fades and the totality ends. Another hour of partial phases remains before the eclipse is over.

While filters are required for viewing the partial phases, they must be removed for totality. The total phase is the only time it is completely safe to look directly at the Sun without protection. In fact, the total phase is not even visible through solar filters because the Sun's corona is a million times fainter than the photosphere[3].

## Figure 1–3: Solar Eclipse Contacts

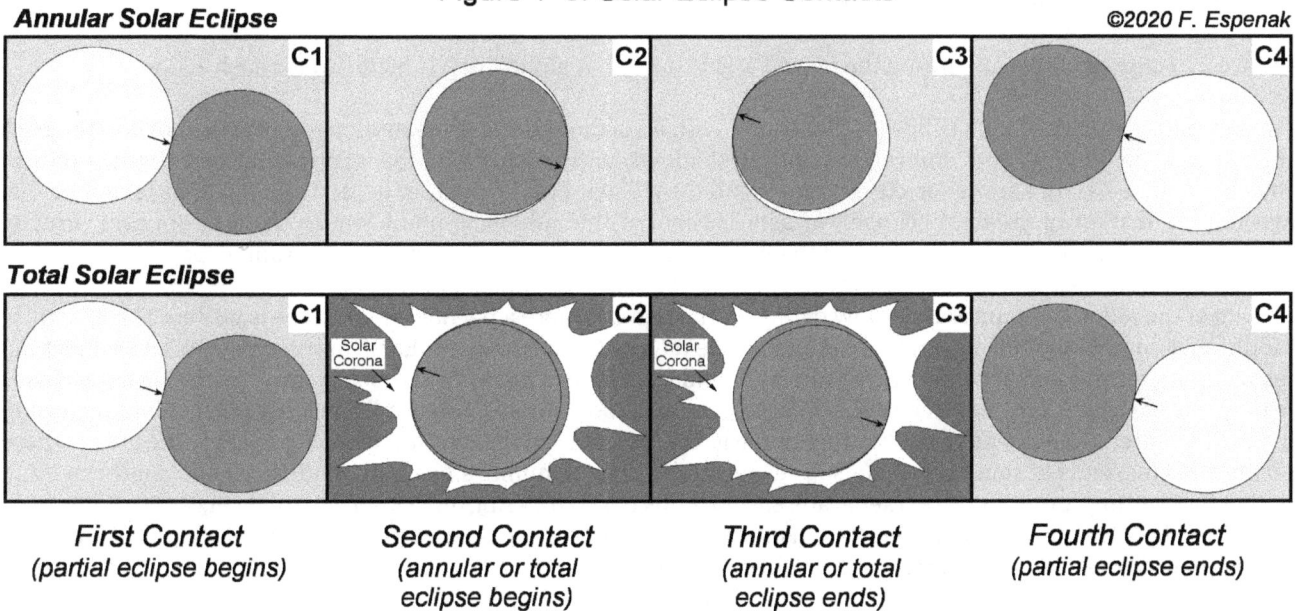

Annular Solar Eclipse

©2020 F. Espenak

Total Solar Eclipse

| First Contact (partial eclipse begins) | Second Contact (annular or total eclipse begins) | Third Contact (annular or total eclipse ends) | Fourth Contact (partial eclipse ends) |

Figure 1–3 illustrates the four contacts for annular and total solar eclipses. The arrows indicate the contact point of the Moon with the Sun's limb in each diagram.

## Solar Eclipse Contacts

During the course of a solar eclipse, the instants when the edge of the Moon's disk becomes tangent to the Sun's disk are known as eclipse contacts. They mark various stages or phases of a solar eclipse (Figure 1–3).

Partial solar eclipses have two primary contacts.

**First Contact (C1) —** Partial Eclipse Begins (Instant of first exterior tangency of the Moon with the Sun)
**Fourth Contact (C4) —** Partial Eclipse Ends (Instant of last exterior tangency of the Moon with the Sun)

---

[2] Shadow bands are caused by the thin solar crescent illuminating Earth's atmosphere before and after totality.
[3] The photosphere is the visible surface of the Sun's disk.

Central solar eclipses (total, annular or hybrid) have four primary contacts. Contacts C2 and C3 mark the instants when the Moon's disk is first and last internally tangent to the Sun. These are the times when the annular or total phase of the eclipse begins and ends, respectively.

**First Contact (C1)** — Partial Eclipse Begins (Instant of first exterior tangency of the Moon with the Sun)
**Second Contact (C2)** — Total (or Annular) Eclipse Begins (Instant of first interior tangency of Moon with Sun)
**Third Contact (C3)** — Total (or Annular) Eclipse Ends (Instant of last interior tangency of Moon with Sun)
**Fourth Contact (C4)** — Partial Eclipse Ends (Instant of last exterior tangency of the Moon with the Sun)

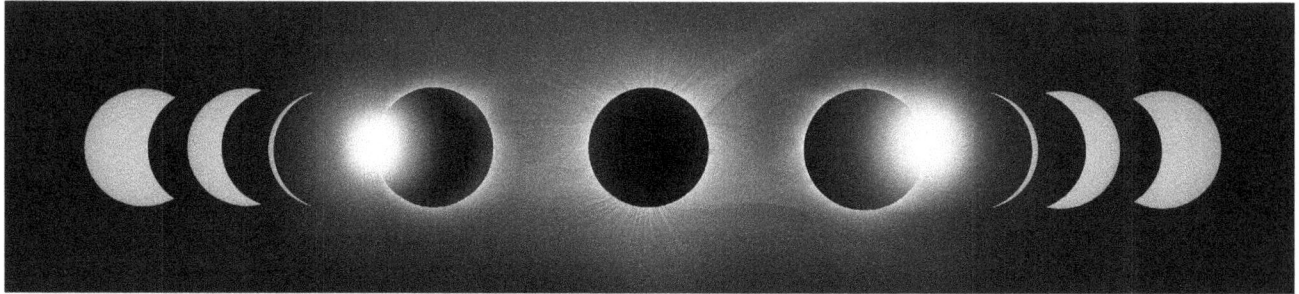

*Photo 1–6 Sequence of partial phases, Diamond Rings, and Totality. Total Solar Eclipse of 2017 Aug 21. ©2017 F. Espenak*

## Explanation of Global Solar Eclipse Maps

There are 22 eclipses of the Sun during the period 2041 to 2050. A global map is included for each eclipse.

The geographic visibility of each eclipse is illustrated with an orthographic projection map of Earth showing the path of the Moon's penumbral and umbral/antumbral shadows with respect to the continental coastlines, political boundaries (circa 2016), and major cities (represented by black dots). North is up and the daylight terminator is drawn for the instant of greatest eclipse. An asterisk marks the sub-solar point where the Sun appears directly overhead at that time. The salient features of the eclipse maps are identified in the key diagram on page 11.

The limits of the Moon's penumbral shadow delineate the region where a partial solar eclipse is visible. This irregular or saddle shaped region often covers more than half the daylight hemisphere of Earth and consists of several distinct zones or limits. At the northern and/or southern boundaries lie the limits of the penumbra's path. Partial eclipses have only one of these limits, as do central eclipses when the Moon's shadow axis falls no closer than about 0.45 radii from Earth's center. Great loops at the western and eastern extremes of the penumbra's path identify the areas where the eclipse begins/ends at sunrise and sunset, respectively. If the penumbra has both a northern and southern limit, the rising and setting curves form two separate, closed loops (e.g., 2024 Apr 08). Otherwise, the curves are connected in a distorted figure eight (e.g., 2021 Dec 04). Bisecting the *eclipse begins/ends at sunrise and sunset* loops is the curve of maximum eclipse at sunrise (western loop) and sunset (eastern loop).

The eclipse magnitude is defined as the fraction of the Sun's diameter occulted by the Moon. A curve of constant eclipse magnitude delineates the points where the local magnitude at maximum eclipse is equal to a constant value. The maps include *curves of constant eclipse magnitude* for values of 0.2, 0.4, 0.6, and 0.8. These curves run exclusively between the curves of maximum eclipse at sunrise and sunset. They are approximately parallel to the northern/southern penumbral limits and the umbral/antumbral paths of central eclipses. The northern and southern limits of the penumbra may be thought of as curves of eclipse magnitude of 0.0. For total eclipses, the northern and southern limits of the umbra are curves of eclipse magnitude of 1.0.

Greatest eclipse is the instant when the axis of the Moon's shadow cone passes closest to Earth's center. The point on Earth's surface intersected by the axis of the Moon's shadow cone at greatest eclipse is marked by an asterisk. For partial eclipses, the shadow axis misses Earth entirely, so the point of greatest eclipse lies on the day/night terminator and the Sun appears on the horizon.

Parameters for the eclipse appears to the right of each map. They include the time of greatest eclipse (Universal Time[4] or UT1), the eclipse magnitude and obscuration, gamma[5], Saros series, and node (ascending or descending).

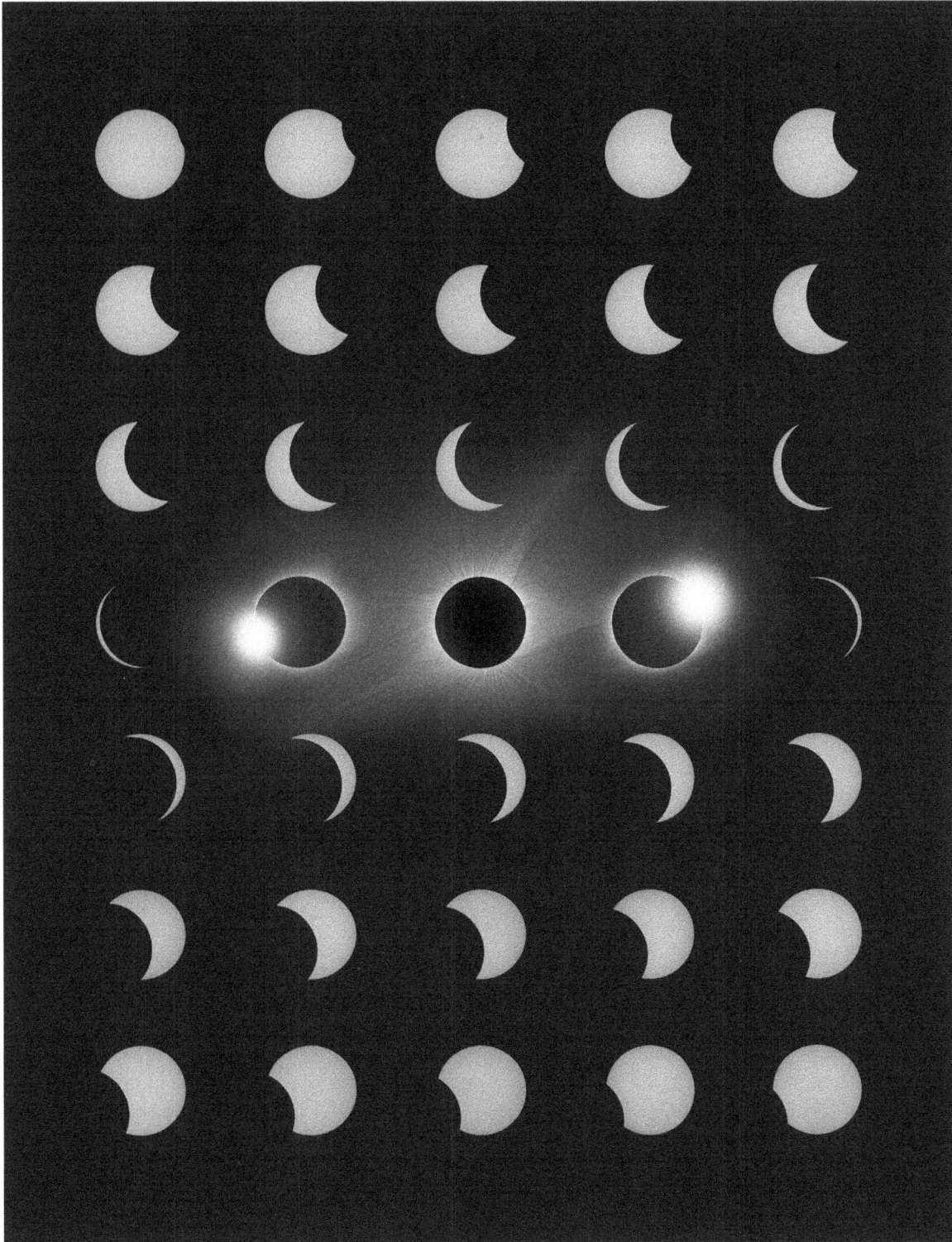

*Photo 1–7 Time Sequence of the Total Solar Eclipse of 2017 Aug 21. ©2017 F. Espenak*

---

[4] Universal Time (UT1) is the modern-day replacement for Greenwich Mean Time.
[5] Gamma is the distance of the Moon's shadow cone axis from Earth's center at greatest eclipse (in Earth equatorial radii).

## Global Solar Eclipse Maps

## Key to Global Solar Eclipse Maps

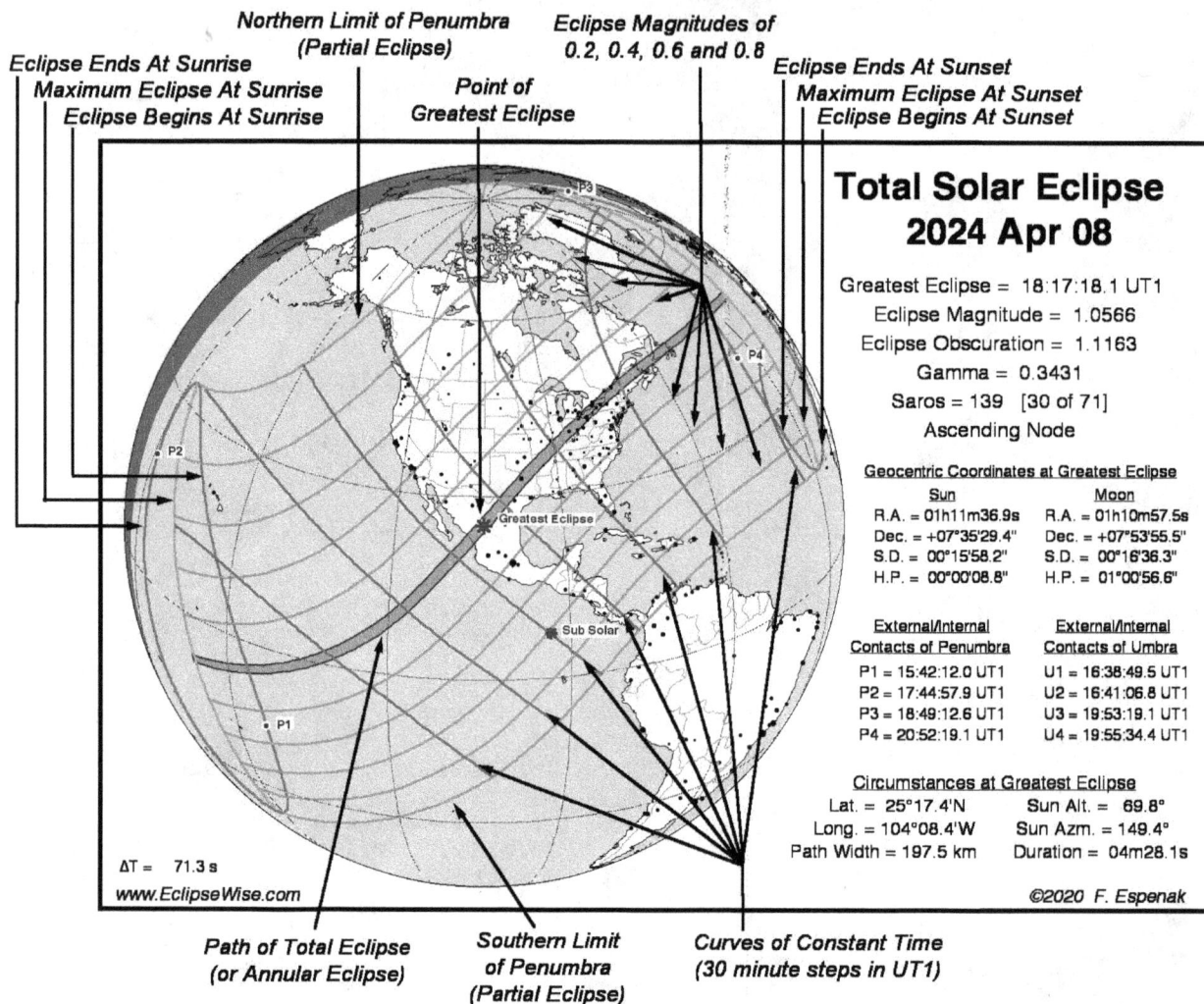

Northern Limit of Penumbra
(Partial Eclipse)

Eclipse Magnitudes of
0.2, 0.4, 0.6 and 0.8

Eclipse Ends At Sunrise
Maximum Eclipse At Sunrise
Eclipse Begins At Sunrise

Point of
Greatest Eclipse

Eclipse Ends At Sunset
Maximum Eclipse At Sunset
Eclipse Begins At Sunset

### Total Solar Eclipse
### 2024 Apr 08

Greatest Eclipse = 18:17:18.1 UT1
Eclipse Magnitude = 1.0566
Eclipse Obscuration = 1.1163
Gamma = 0.3431
Saros = 139  [30 of 71]
Ascending Node

**Geocentric Coordinates at Greatest Eclipse**

| Sun | Moon |
|---|---|
| R.A. = 01h11m36.9s | R.A. = 01h10m57.5s |
| Dec. = +07°35'29.4" | Dec. = +07°53'55.5" |
| S.D. = 00°15'58.2" | S.D. = 00°16'36.3" |
| H.P. = 00°00'08.8" | H.P. = 01°00'56.6" |

| External/Internal Contacts of Penumbra | External/Internal Contacts of Umbra |
|---|---|
| P1 = 15:42:12.0 UT1 | U1 = 16:38:49.5 UT1 |
| P2 = 17:44:57.9 UT1 | U2 = 16:41:06.8 UT1 |
| P3 = 18:49:12.6 UT1 | U3 = 19:53:19.1 UT1 |
| P4 = 20:52:19.1 UT1 | U4 = 19:55:34.4 UT1 |

**Circumstances at Greatest Eclipse**

| | |
|---|---|
| Lat. = 25°17.4'N | Sun Alt. = 69.8° |
| Long. = 104°08.4'W | Sun Azm. = 149.4° |
| Path Width = 197.5 km | Duration = 04m28.1s |

ΔT = 71.3 s
www.EclipseWise.com

©2020 F. Espenak

Path of Total Eclipse
(or Annular Eclipse)

Southern Limit
of Penumbra
(Partial Eclipse)

Curves of Constant Time
(30 minute steps in UT1)

## Explanation of Terms Used in Global Solar Eclipse Maps

**Greatest Eclipse** – The instant when the distance between the axis of the Moon's shadow cone and the center of Earth reaches a minimum (in Universal Time[6] or UT1).

**Eclipse Magnitude** – The fraction of the Sun's diameter occulted by the Moon at the instant of greatest eclipse (for total and annular eclipses this value is the ratio of diameters of the Moon and the Sun).

**Eclipse Obscuration** – The fraction of the Sun's area occulted by the Moon at the instant of greatest eclipse.

**Gamma** – The minimum distance from the lunar shadow axis to the center of Earth (units of Earth equatorial radii).

**Saros Series** – The Saros series that the eclipse belongs to. The numbers in "[ ]" are the eclipse's sequential position and the number of eclipses in the Saros series.

**Node** – The orbital node near which the eclipse takes place (Ascending Node or Descending Node).

**Geocentric Coordinates of the Sun and the Moon at Greatest Eclipse**

| | |
|---|---|
| **R.A.** – Right Ascension | **S.D.** – Semi-Diameter (i.e. - radius) |
| **Dec.** – Declination | **H.P.** – Horizontal Parallax |

**External/Internal Contacts of Penumbra and Umbra** – Instants when each shadow enters or exits the surface of Earth (Penumbral contacts shown on map as: P1, P2, P3, and P4; umbral contacts are located at the ends of the central path). All times are in Universal Time (UT1)

**Circumstances of Greatest Eclipse** – Geographic location (Lat., Long.) of the shadow axis, the altitude and azimuth of the Sun, width of the central path, and the central duration of totality or annularity.

---

[6] Universal Time or UT1 is the modern replacement for Greenwich Mean Time

# Total Solar Eclipse
# 2041 Apr 30

Greatest Eclipse = 11:51:00.9 UT1

Eclipse Magnitude = 1.0189

Eclipse Obscuration = 1.0382

Gamma = -0.4492

Saros = 129   [53 of 80]

Ascending Node

### Geocentric Coordinates at Greatest Eclipse

| Sun | Moon |
|---|---|
| R.A. = 02h32m22.2s | R.A. = 02h33m06.0s |
| Dec. = +14°58'18.8" | Dec. = +14°34'20.1" |
| S.D. = 00°15'52.6" | S.D. = 00°15'56.6" |
| H.P. = 00°00'08.7" | H.P. = 00°58'30.8" |

| External/Internal Contacts of Penumbra | External/Internal Contacts of Umbra |
|---|---|
| P1 = 09:11:07.3 UT1 | U1 = 10:14:21.9 UT1 |
| P2 = 11:45:36.7 UT1 | U2 = 10:14:38.7 UT1 |
| P3 = 11:56:54.1 UT1 | U3 = 13:27:40.0 UT1 |
| P4 = 14:31:09.0 UT1 | U4 = 13:27:51.4 UT1 |

### Circumstances at Greatest Eclipse

| | |
|---|---|
| Lat. = 09°37.2'S | Sun Alt. = 63.2° |
| Long. = 012°09.9'E | Sun Azm. = 336.7° |
| Path Width = 71.7 km | Duration = 01m50.6s |

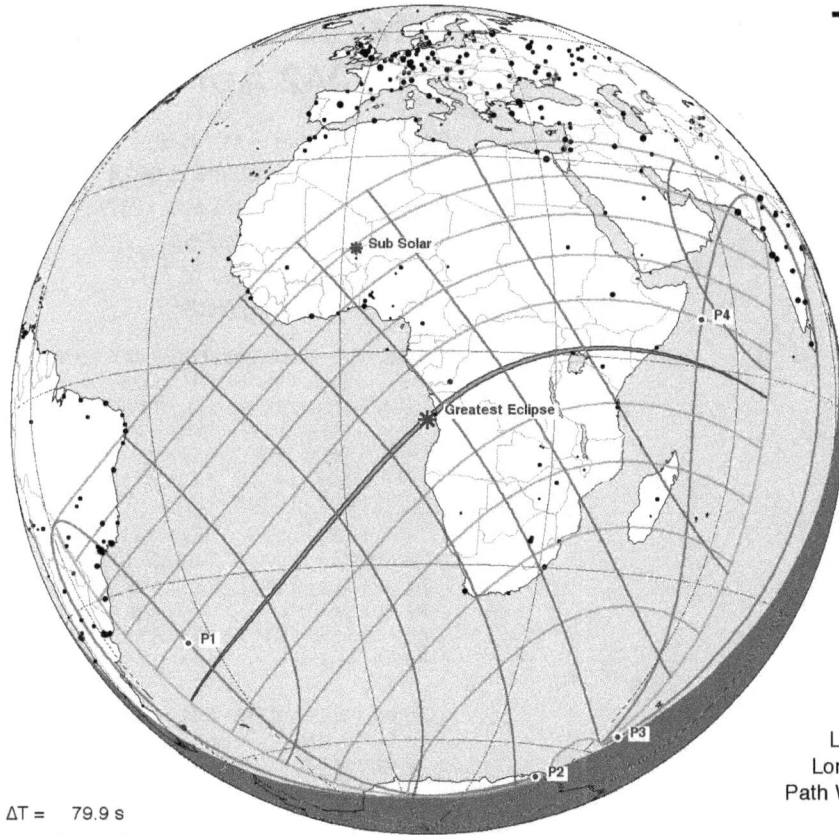

ΔT =   79.9 s

www.EclipseWise.com

©2020  F. Espenak

# Annular Solar Eclipse
# 2041 Oct 25

Greatest Eclipse = 01:35:01.6 UT1

Eclipse Magnitude = 0.9467

Eclipse Obscuration = 0.8962

Gamma = 0.4133

Saros = 134   [45 of 71]

Decending Node

### Geocentric Coordinates at Greatest Eclipse

| Sun | Moon |
|---|---|
| R.A. = 13h59m22.0s | R.A. = 14h00m02.5s |
| Dec. = -12°10'20.1" | Dec. = -11°49'54.3" |
| S.D. = 00°16'04.9" | S.D. = 00°15'00.8" |
| H.P. = 00°00'08.8" | H.P. = 00°55'06.0" |

| External/Internal Contacts of Penumbra | External/Internal Contacts of Umbra |
|---|---|
| P1 = 22:39:42.7 UT1 | U1 = 23:46:58.4 UT1 |
| P2 = 01:20:51.1 UT1 | U2 = 23:52:01.6 UT1 |
| P3 = 01:49:46.9 UT1 | U3 = 03:18:18.2 UT1 |
| P4 = 04:30:24.6 UT1 | U4 = 03:23:16.1 UT1 |

### Circumstances at Greatest Eclipse

| | |
|---|---|
| Lat. = 09°56.0'N | Sun Alt. = 65.5° |
| Long. = 162°50.0'E | Sun Azm. = 205.6° |
| Path Width = 213.1 km | Duration = 06m07.1s |

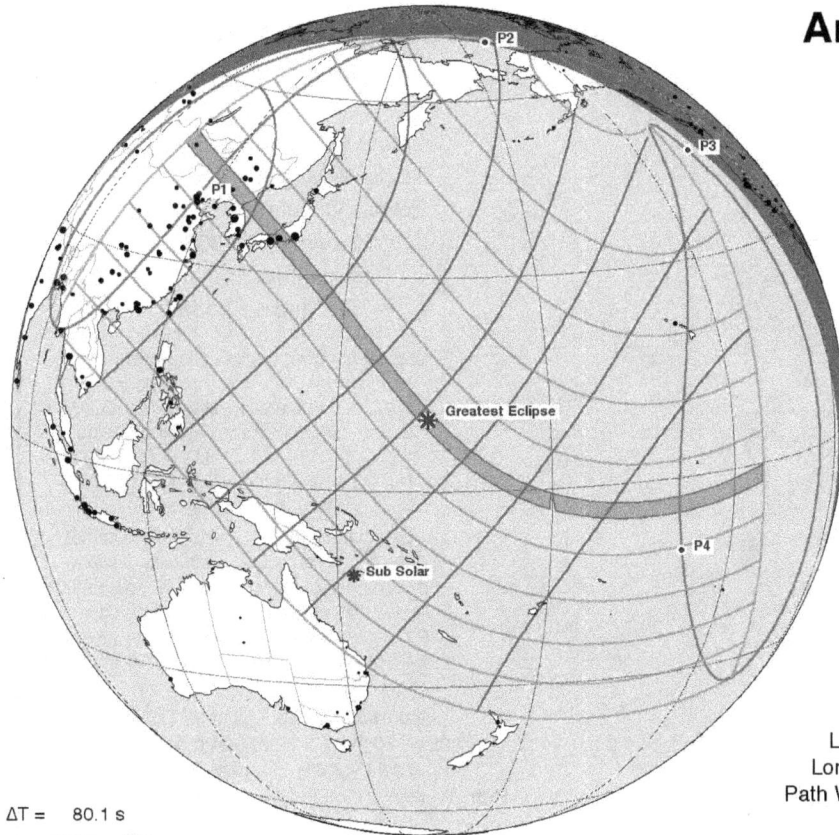

ΔT =   80.1 s

www.EclipseWise.com

©2020  F. Espenak

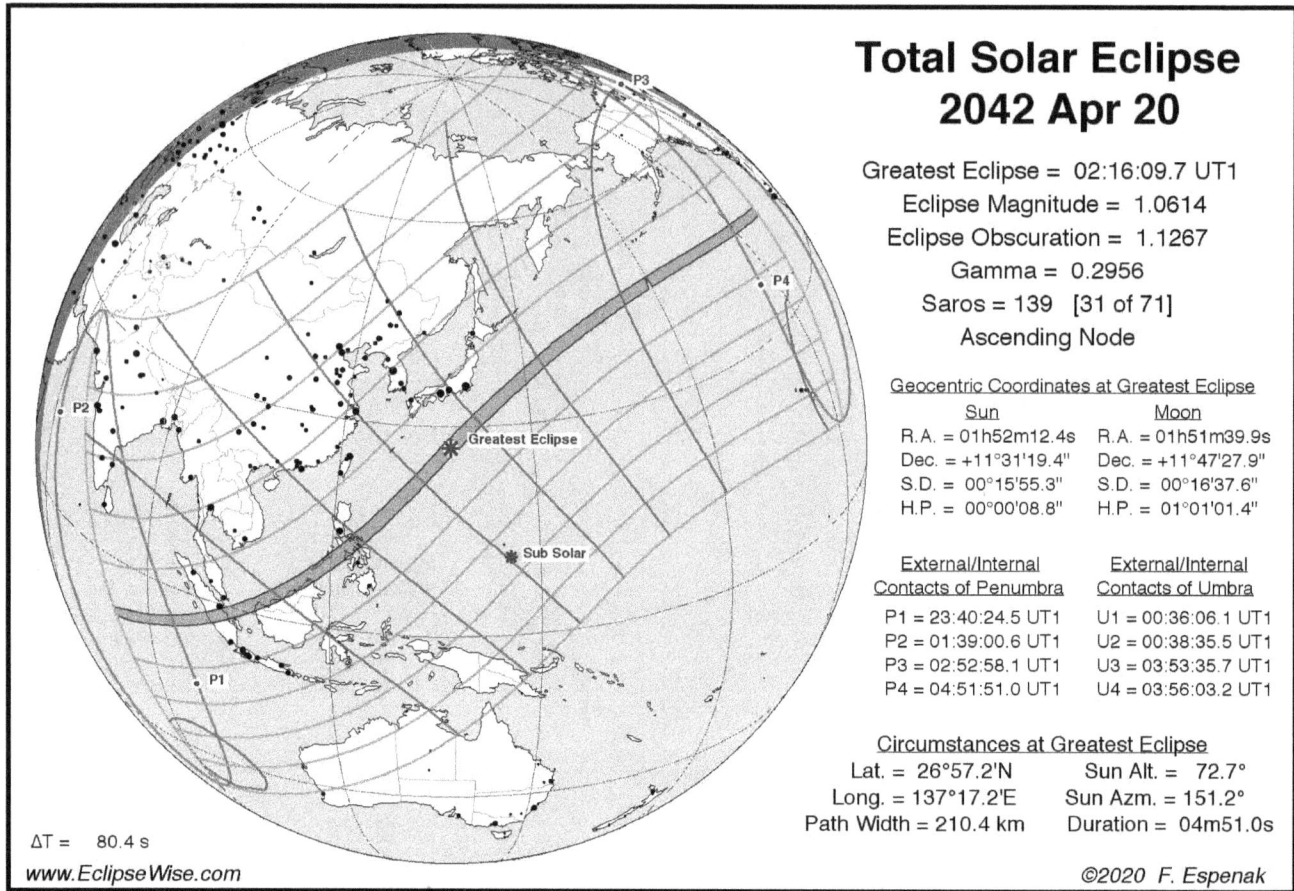

# Total Solar Eclipse
# 2042 Apr 20

Greatest Eclipse = 02:16:09.7 UT1

Eclipse Magnitude = 1.0614

Eclipse Obscuration = 1.1267

Gamma = 0.2956

Saros = 139 [31 of 71]

Ascending Node

### Geocentric Coordinates at Greatest Eclipse

| | Sun | Moon |
|---|---|---|
| R.A. = | 01h52m12.4s | R.A. = 01h51m39.9s |
| Dec. = | +11°31'19.4" | Dec. = +11°47'27.9" |
| S.D. = | 00°15'55.3" | S.D. = 00°16'37.6" |
| H.P. = | 00°00'08.8" | H.P. = 01°01'01.4" |

| External/Internal Contacts of Penumbra | External/Internal Contacts of Umbra |
|---|---|
| P1 = 23:40:24.5 UT1 | U1 = 00:36:06.1 UT1 |
| P2 = 01:39:00.6 UT1 | U2 = 00:38:35.5 UT1 |
| P3 = 02:52:58.1 UT1 | U3 = 03:53:35.7 UT1 |
| P4 = 04:51:51.0 UT1 | U4 = 03:56:03.2 UT1 |

### Circumstances at Greatest Eclipse

| | |
|---|---|
| Lat. = 26°57.2'N | Sun Alt. = 72.7° |
| Long. = 137°17.2'E | Sun Azm. = 151.2° |
| Path Width = 210.4 km | Duration = 04m51.0s |

ΔT = 80.4 s

www.EclipseWise.com

©2020 F. Espenak

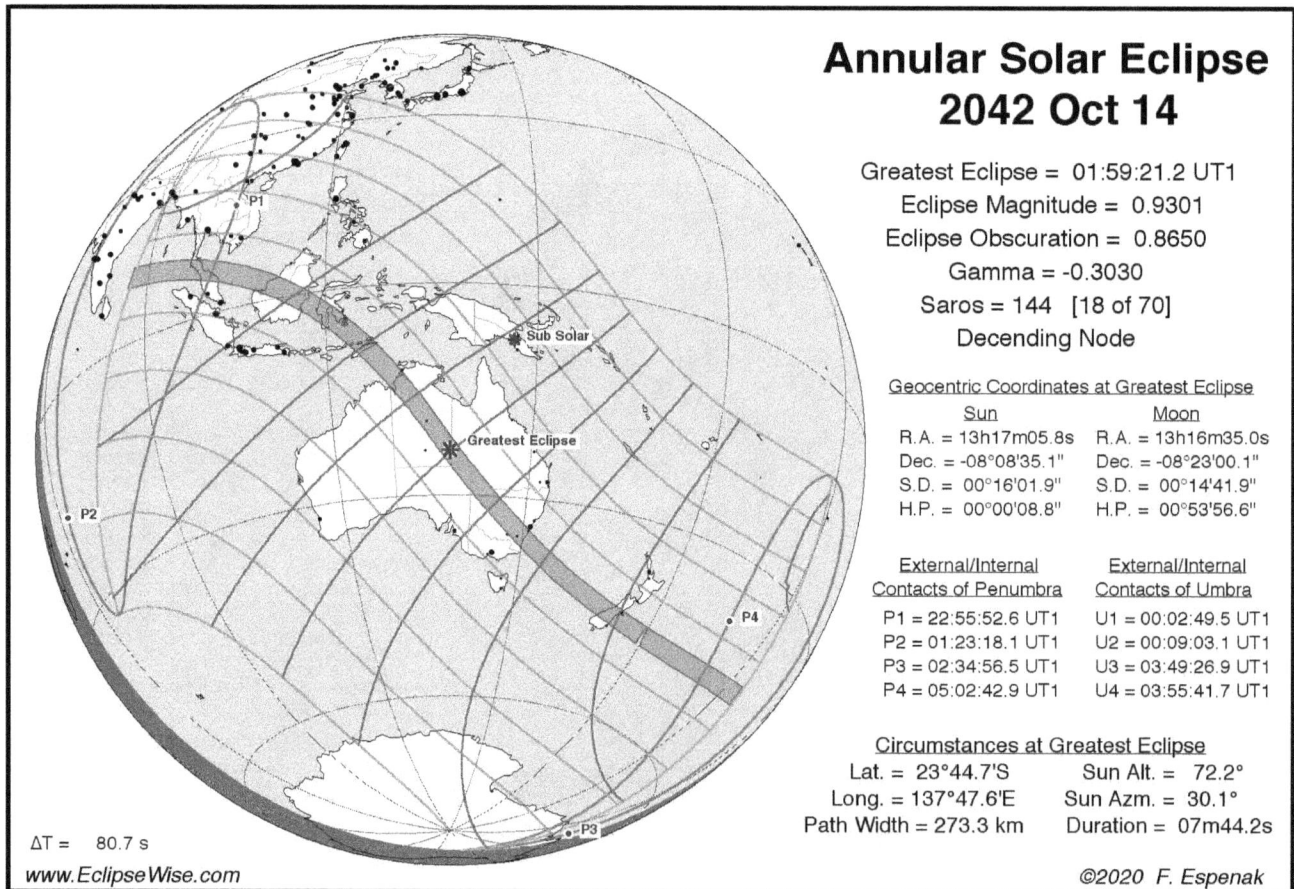

# Annular Solar Eclipse
# 2042 Oct 14

Greatest Eclipse = 01:59:21.2 UT1

Eclipse Magnitude = 0.9301

Eclipse Obscuration = 0.8650

Gamma = -0.3030

Saros = 144 [18 of 70]

Decending Node

### Geocentric Coordinates at Greatest Eclipse

| | Sun | Moon |
|---|---|---|
| R.A. = | 13h17m05.8s | R.A. = 13h16m35.0s |
| Dec. = | -08°08'35.1" | Dec. = -08°23'00.1" |
| S.D. = | 00°16'01.9" | S.D. = 00°14'41.9" |
| H.P. = | 00°00'08.8" | H.P. = 00°53'56.6" |

| External/Internal Contacts of Penumbra | External/Internal Contacts of Umbra |
|---|---|
| P1 = 22:55:52.6 UT1 | U1 = 00:02:49.5 UT1 |
| P2 = 01:23:18.1 UT1 | U2 = 00:09:03.1 UT1 |
| P3 = 02:34:56.5 UT1 | U3 = 03:49:26.9 UT1 |
| P4 = 05:02:42.9 UT1 | U4 = 03:55:41.7 UT1 |

### Circumstances at Greatest Eclipse

| | |
|---|---|
| Lat. = 23°44.7'S | Sun Alt. = 72.2° |
| Long. = 137°47.6'E | Sun Azm. = 30.1° |
| Path Width = 273.3 km | Duration = 07m44.2s |

ΔT = 80.7 s

www.EclipseWise.com

©2020 F. Espenak

# Total Solar Eclipse
# 2043 Apr 09

Greatest Eclipse = 18:56:28.4 UT1

Eclipse Magnitude = 1.0096

Eclipse Obscuration =-NAN(001)

Gamma = 1.0031

Saros = 149 [22 of 71]

Ascending Node

### Geocentric Coordinates at Greatest Eclipse

| Sun | Moon |
|---|---|
| R.A. = 01h13m12.2s | R.A. = 01h11m17.3s |
| Dec. = +07°45'05.1" | Dec. = +08°39'09.1" |
| S.D. = 00°15'58.1" | S.D. = 00°16'38.0" |
| H.P. = 00°00'08.8" | H.P. = 01°01'02.7" |

| External/Internal Contacts of Penumbra | External/Internal Contacts of Umbra |
|---|---|
| P1 = 16:56:13.2 UT1 | U1 = 18:45:47.3 UT1 |
| P4 = 20:56:19.4 UT1 | U4 = 19:06:37.1 UT1 |

### (Non-Central)
### Circumstances at Greatest Eclipse

| | |
|---|---|
| Lat. = 61°18.7'N | Sun Alt. = 0.0° |
| Long. = 151°56.3'E | Sun Azm. = 73.7° |
| Path Width = 0.0 km | Duration = 01m46.3s |

ΔT = 81.0 s

www.EclipseWise.com

©2020 F. Espenak

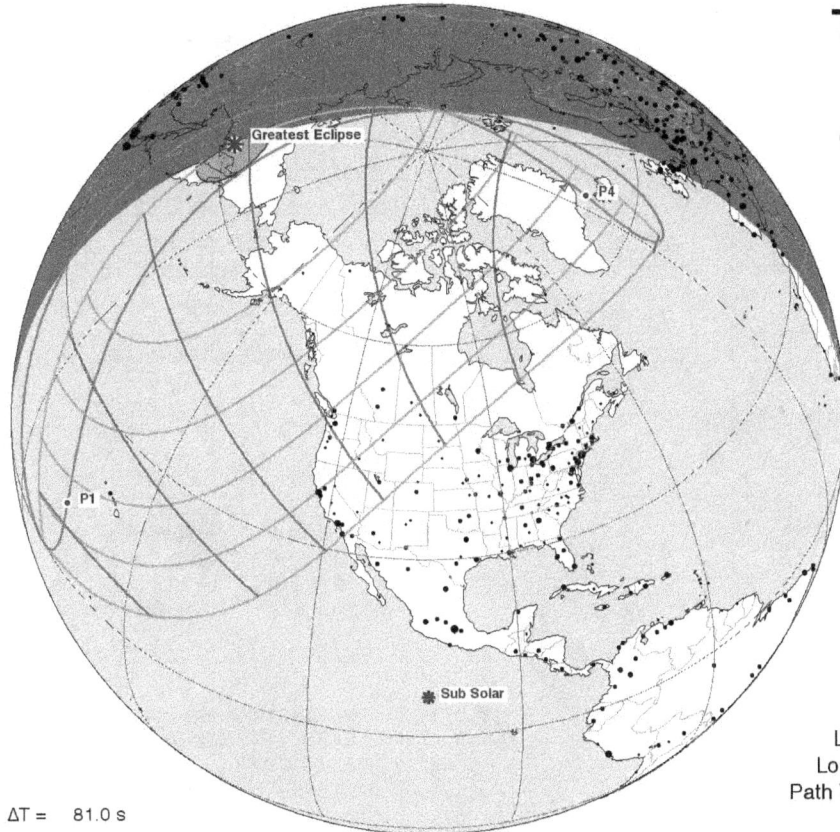

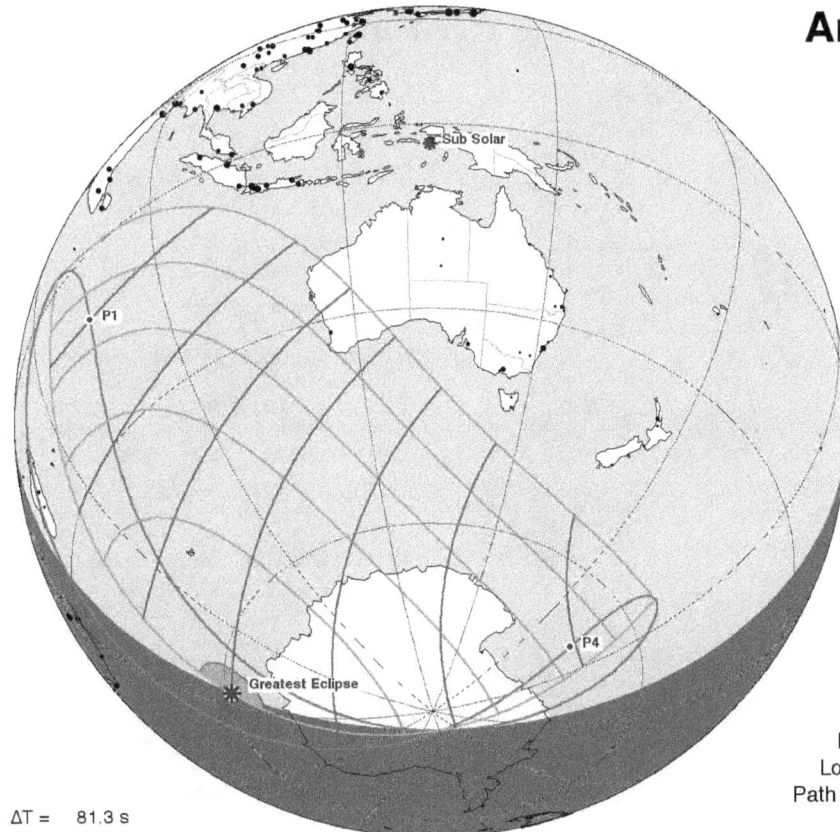

# Annular Solar Eclipse
# 2043 Oct 03

Greatest Eclipse = 03:00:27.7 UT1

Eclipse Magnitude = 0.9497

Eclipse Obscuration =-NAN(001)

Gamma = -1.0102

Saros = 154 [8 of 71]

Decending Node

### Geocentric Coordinates at Greatest Eclipse

| Sun | Moon |
|---|---|
| R.A. = 12h36m02.9s | R.A. = 12h34m15.0s |
| Dec. = -03°53'04.6" | Dec. = -04°41'56.9" |
| S.D. = 00°15'58.8" | S.D. = 00°15'05.1" |
| H.P. = 00°00'08.8" | H.P. = 00°55'21.7" |

| External/Internal Contacts of Penumbra | External/Internal Contacts of Umbra |
|---|---|
| P1 = 00:42:55.5 UT1 | U1 = 02:50:16.1 UT1 |
| P4 = 05:17:39.8 UT1 | U4 = 03:10:03.1 UT1 |

### (Non-Central)
### Circumstances at Greatest Eclipse

| | |
|---|---|
| Lat. = 60°57.6'S | Sun Alt. = 0.0° |
| Long. = 035°14.2'E | Sun Azm. = 98.0° |
| Path Width = 0.0 km | Duration = 02m17.0s |

ΔT = 81.3 s

www.EclipseWise.com

©2020 F. Espenak

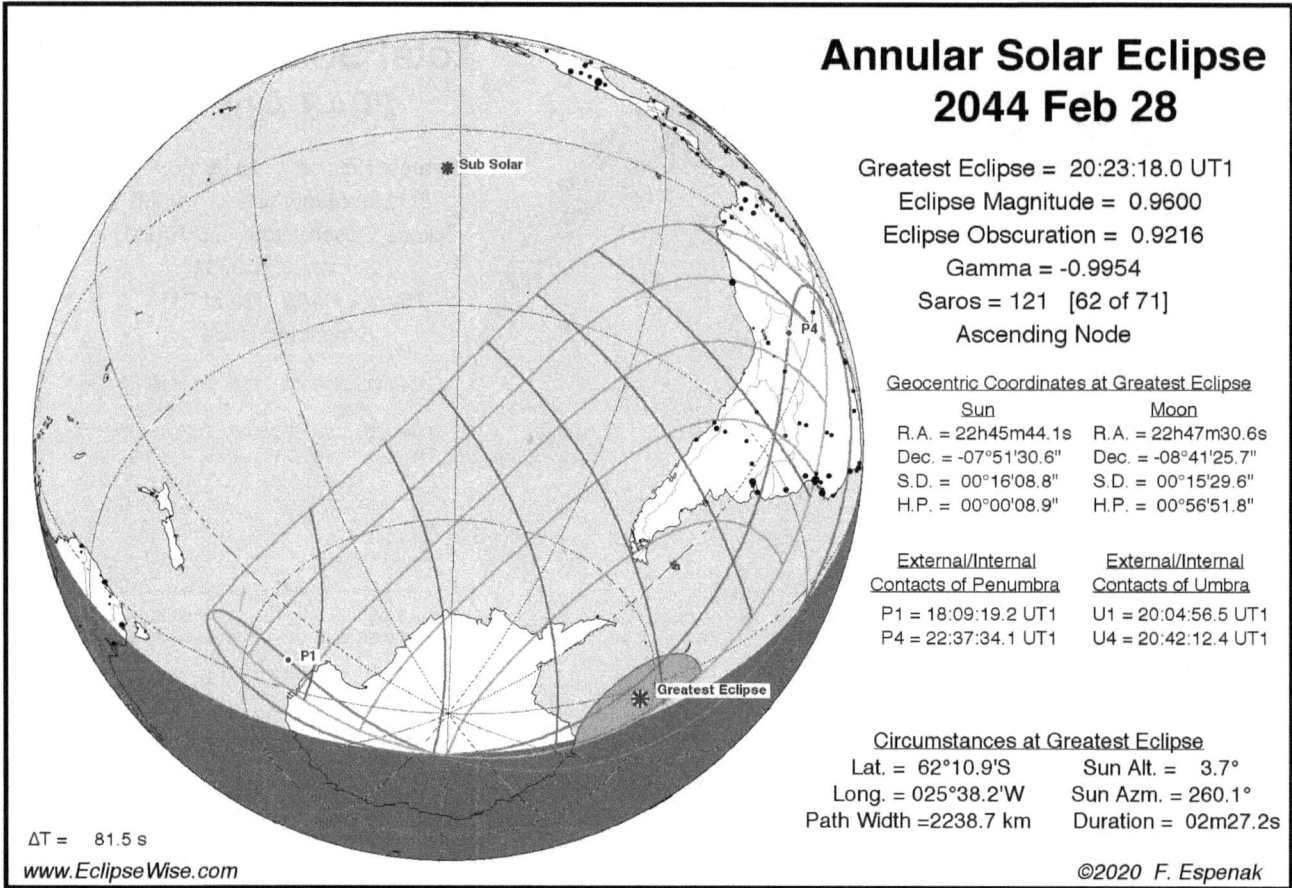

# Annular Solar Eclipse
# 2044 Feb 28

Greatest Eclipse = 20:23:18.0 UT1

Eclipse Magnitude = 0.9600

Eclipse Obscuration = 0.9216

Gamma = -0.9954

Saros = 121  [62 of 71]

Ascending Node

### Geocentric Coordinates at Greatest Eclipse

| Sun | Moon |
|---|---|
| R.A. = 22h45m44.1s | R.A. = 22h47m30.6s |
| Dec. = -07°51'30.6" | Dec. = -08°41'25.7" |
| S.D. = 00°16'08.8" | S.D. = 00°15'29.6" |
| H.P. = 00°00'08.9" | H.P. = 00°56'51.8" |

| External/Internal Contacts of Penumbra | External/Internal Contacts of Umbra |
|---|---|
| P1 = 18:09:19.2 UT1 | U1 = 20:04:56.5 UT1 |
| P4 = 22:37:34.1 UT1 | U4 = 20:42:12.4 UT1 |

### Circumstances at Greatest Eclipse

| | |
|---|---|
| Lat. = 62°10.9'S | Sun Alt. = 3.7° |
| Long. = 025°38.2'W | Sun Azm. = 260.1° |
| Path Width =2238.7 km | Duration = 02m27.2s |

ΔT = 81.5 s

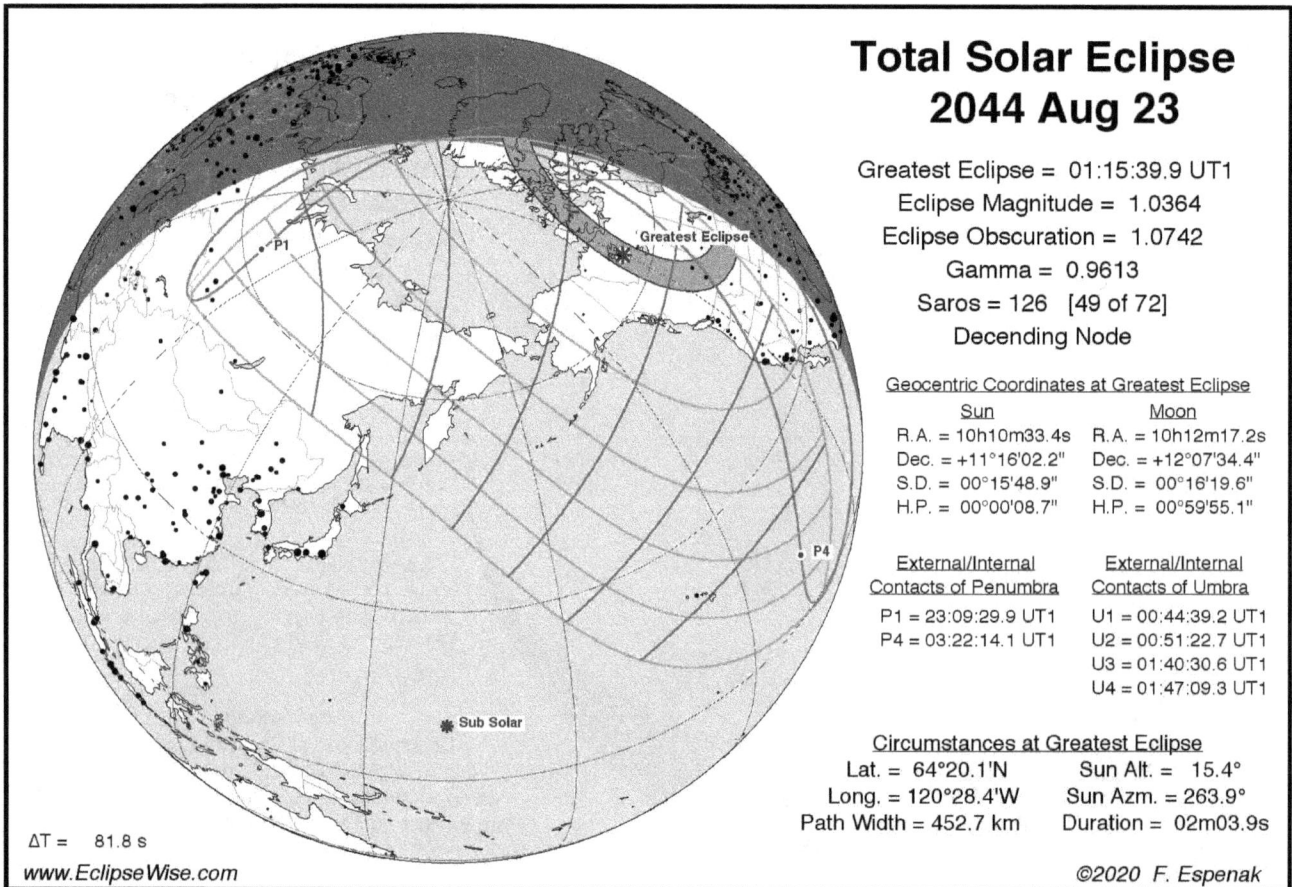

# Total Solar Eclipse
# 2044 Aug 23

Greatest Eclipse = 01:15:39.9 UT1

Eclipse Magnitude = 1.0364

Eclipse Obscuration = 1.0742

Gamma = 0.9613

Saros = 126  [49 of 72]

Decending Node

### Geocentric Coordinates at Greatest Eclipse

| Sun | Moon |
|---|---|
| R.A. = 10h10m33.4s | R.A. = 10h12m17.2s |
| Dec. = +11°16'02.2" | Dec. = +12°07'34.4" |
| S.D. = 00°15'48.9" | S.D. = 00°16'19.6" |
| H.P. = 00°00'08.7" | H.P. = 00°59'55.1" |

| External/Internal Contacts of Penumbra | External/Internal Contacts of Umbra |
|---|---|
| P1 = 23:09:29.9 UT1 | U1 = 00:44:39.2 UT1 |
| P4 = 03:22:14.1 UT1 | U2 = 00:51:22.7 UT1 |
| | U3 = 01:40:30.6 UT1 |
| | U4 = 01:47:09.3 UT1 |

### Circumstances at Greatest Eclipse

| | |
|---|---|
| Lat. = 64°20.1'N | Sun Alt. = 15.4° |
| Long. = 120°28.4'W | Sun Azm. = 263.9° |
| Path Width = 452.7 km | Duration = 02m03.9s |

ΔT = 81.8 s

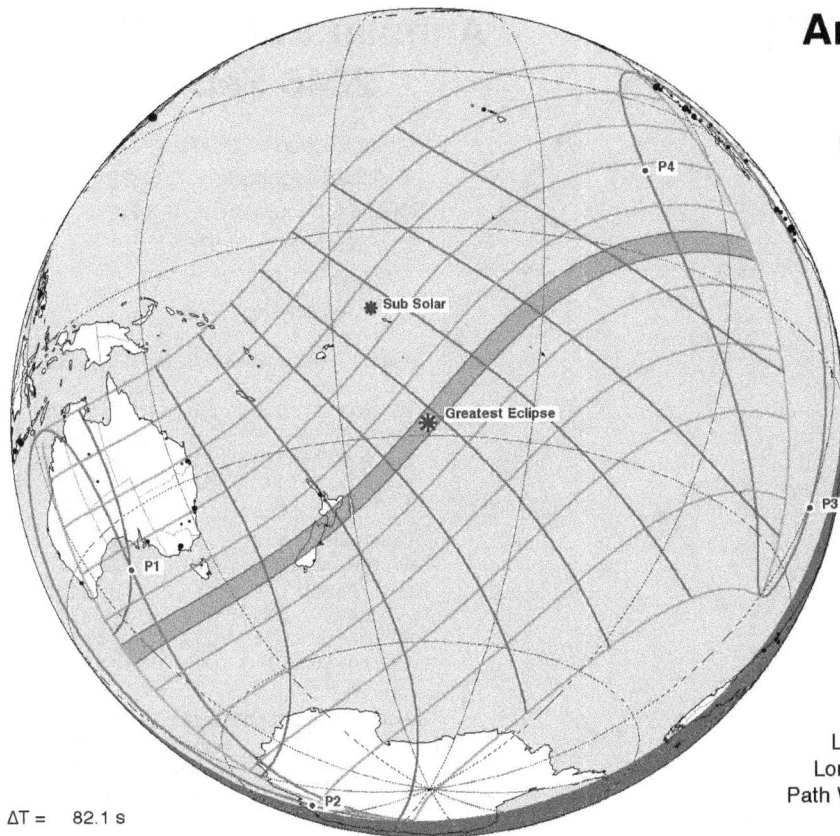

# Annular Solar Eclipse
## 2045 Feb 16

Greatest Eclipse = 23:54:44.5 UT1

Eclipse Magnitude = 0.9285

Eclipse Obscuration = 0.8621

Gamma = -0.3125

Saros = 131  [52 of 70]

Ascending Node

### Geocentric Coordinates at Greatest Eclipse

| Sun | Moon |
|---|---|
| R.A. = 22h03m27.1s | R.A. = 22h03m57.6s |
| Dec. = -11°55'04.8" | Dec. = -12°10'17.7" |
| S.D. = 00°16'11.2" | S.D. = 00°14'48.9" |
| H.P. = 00°00'08.9" | H.P. = 00°54'22.2" |

| External/Internal Contacts of Penumbra | External/Internal Contacts of Umbra |
|---|---|
| P1 = 20:52:55.0 UT1 | U1 = 21:59:33.6 UT1 |
| P2 = 23:20:41.3 UT1 | U2 = 22:05:56.0 UT1 |
| P3 = 00:29:14.8 UT1 | U3 = 01:43:45.9 UT1 |
| P4 = 02:56:37.3 UT1 | U4 = 01:50:04.2 UT1 |

### Circumstances at Greatest Eclipse

| | |
|---|---|
| Lat. = 28°15.4'S | Sun Alt. = 71.6° |
| Long. = 166°13.6'W | Sun Azm. = 331.0° |
| Path Width = 281.2 km | Duration = 07m46.5s |

ΔT = 82.1 s

www.EclipseWise.com

©2020  F. Espenak

# Total Solar Eclipse
## 2045 Aug 12

Greatest Eclipse = 17:41:16.7 UT1

Eclipse Magnitude = 1.0774

Eclipse Obscuration = 1.1607

Gamma = 0.2116

Saros = 136  [39 of 71]

Decending Node

### Geocentric Coordinates at Greatest Eclipse

| Sun | Moon |
|---|---|
| R.A. = 09h31m17.7s | R.A. = 09h31m39.7s |
| Dec. = +14°40'40.5" | Dec. = +14°52'29.9" |
| S.D. = 00°15'47.0" | S.D. = 00°16'43.3" |
| H.P. = 00°00'08.7" | H.P. = 01°01'22.3" |

| External/Internal Contacts of Penumbra | External/Internal Contacts of Umbra |
|---|---|
| P1 = 15:05:38.4 UT1 | U1 = 15:59:25.2 UT1 |
| P2 = 16:58:30.3 UT1 | U2 = 16:02:36.5 UT1 |
| P3 = 18:24:16.1 UT1 | U3 = 19:20:03.1 UT1 |
| P4 = 20:16:59.1 UT1 | U4 = 19:23:14.1 UT1 |

### Circumstances at Greatest Eclipse

| | |
|---|---|
| Lat. = 25°54.6'N | Sun Alt. = 77.6° |
| Long. = 078°33.9'W | Sun Azm. = 205.7° |
| Path Width = 255.6 km | Duration = 06m05.7s |

ΔT = 82.4 s

www.EclipseWise.com

©2020  F. Espenak

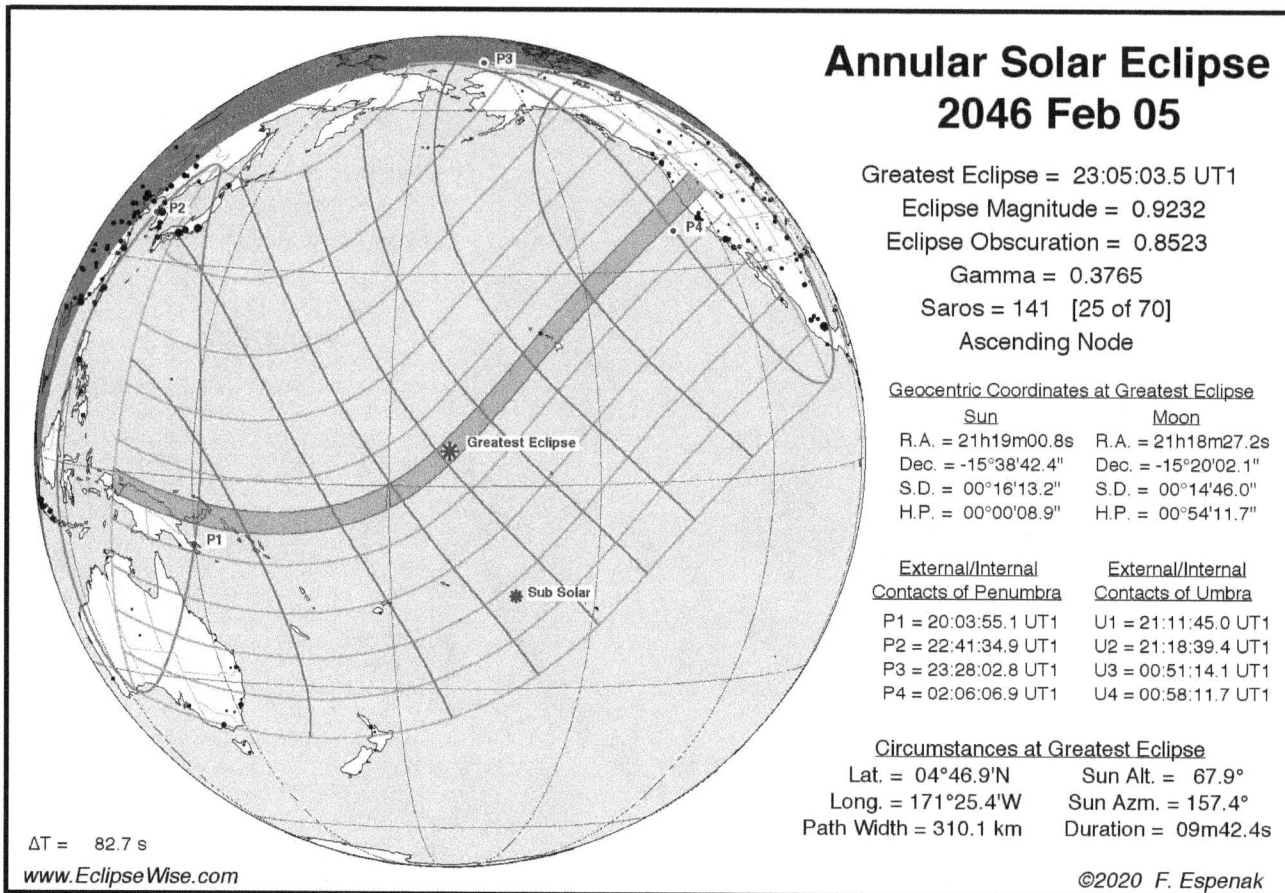

## Annular Solar Eclipse
## 2046 Feb 05

Greatest Eclipse = 23:05:03.5 UT1

Eclipse Magnitude = 0.9232

Eclipse Obscuration = 0.8523

Gamma = 0.3765

Saros = 141 [25 of 70]

Ascending Node

Geocentric Coordinates at Greatest Eclipse

| | Sun | Moon |
|---|---|---|
| R.A. = | 21h19m00.8s | 21h18m27.2s |
| Dec. = | -15°38'42.4" | -15°20'02.1" |
| S.D. = | 00°16'13.2" | 00°14'46.0" |
| H.P. = | 00°00'08.9" | 00°54'11.7" |

| External/Internal Contacts of Penumbra | External/Internal Contacts of Umbra |
|---|---|
| P1 = 20:03:55.1 UT1 | U1 = 21:11:45.0 UT1 |
| P2 = 22:41:34.9 UT1 | U2 = 21:18:39.4 UT1 |
| P3 = 23:28:02.8 UT1 | U3 = 00:51:14.1 UT1 |
| P4 = 02:06:06.9 UT1 | U4 = 00:58:11.7 UT1 |

Circumstances at Greatest Eclipse

| | |
|---|---|
| Lat. = 04°46.9'N | Sun Alt. = 67.9° |
| Long. = 171°25.4'W | Sun Azm. = 157.4° |
| Path Width = 310.1 km | Duration = 09m42.4s |

ΔT = 82.7 s

www.EclipseWise.com

©2020 F. Espenak

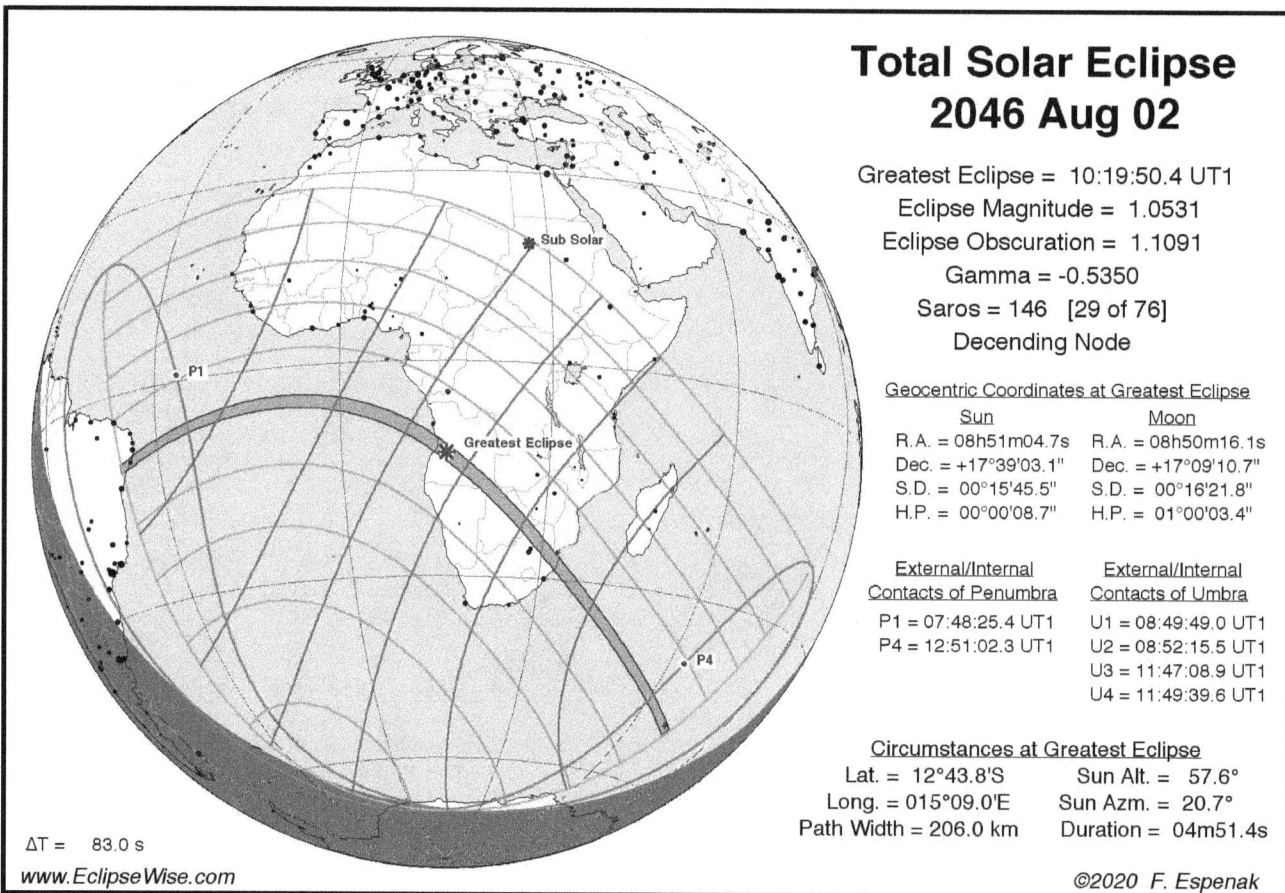

## Total Solar Eclipse
## 2046 Aug 02

Greatest Eclipse = 10:19:50.4 UT1

Eclipse Magnitude = 1.0531

Eclipse Obscuration = 1.1091

Gamma = -0.5350

Saros = 146 [29 of 76]

Decending Node

Geocentric Coordinates at Greatest Eclipse

| | Sun | Moon |
|---|---|---|
| R.A. = | 08h51m04.7s | 08h50m16.1s |
| Dec. = | +17°39'03.1" | +17°09'10.7" |
| S.D. = | 00°15'45.5" | 00°16'21.8" |
| H.P. = | 00°00'08.7" | 01°00'03.4" |

| External/Internal Contacts of Penumbra | External/Internal Contacts of Umbra |
|---|---|
| P1 = 07:48:25.4 UT1 | U1 = 08:49:49.0 UT1 |
| P4 = 12:51:02.3 UT1 | U2 = 08:52:15.5 UT1 |
| | U3 = 11:47:08.9 UT1 |
| | U4 = 11:49:39.6 UT1 |

Circumstances at Greatest Eclipse

| | |
|---|---|
| Lat. = 12°43.8'S | Sun Alt. = 57.6° |
| Long. = 015°09.0'E | Sun Azm. = 20.7° |
| Path Width = 206.0 km | Duration = 04m51.4s |

ΔT = 83.0 s

www.EclipseWise.com

©2020 F. Espenak

# Partial Solar Eclipse
## 2047 Jan 26

Greatest Eclipse = 01:31:54.5 UT1
Eclipse Magnitude = 0.8908
Eclipse Obscuration = 0.8404
Gamma = 1.0450
Saros = 151   [16 of 72]
Ascending Node

### Geocentric Coordinates at Greatest Eclipse

| Sun | Moon |
|---|---|
| R.A. = 20h33m28.4s | R.A. = 20h32m04.0s |
| Dec. = -18°46'10.9" | Dec. = -17°50'50.8" |
| S.D. = 00°16'14.7" | S.D. = 00°15'23.2" |
| H.P. = 00°00'08.9" | H.P. = 00°56'28.0" |

### External/Internal Contacts of Penumbra

P1 = 23:20:46.1 UT1
P4 = 03:42:51.1 UT1

### Circumstances at Greatest Eclipse

| Lat. = 62°52.0'N | Sun Alt. =   0.0° |
|---|---|
| Long. = 111°42.5'E | Sun Azm. = 134.9° |

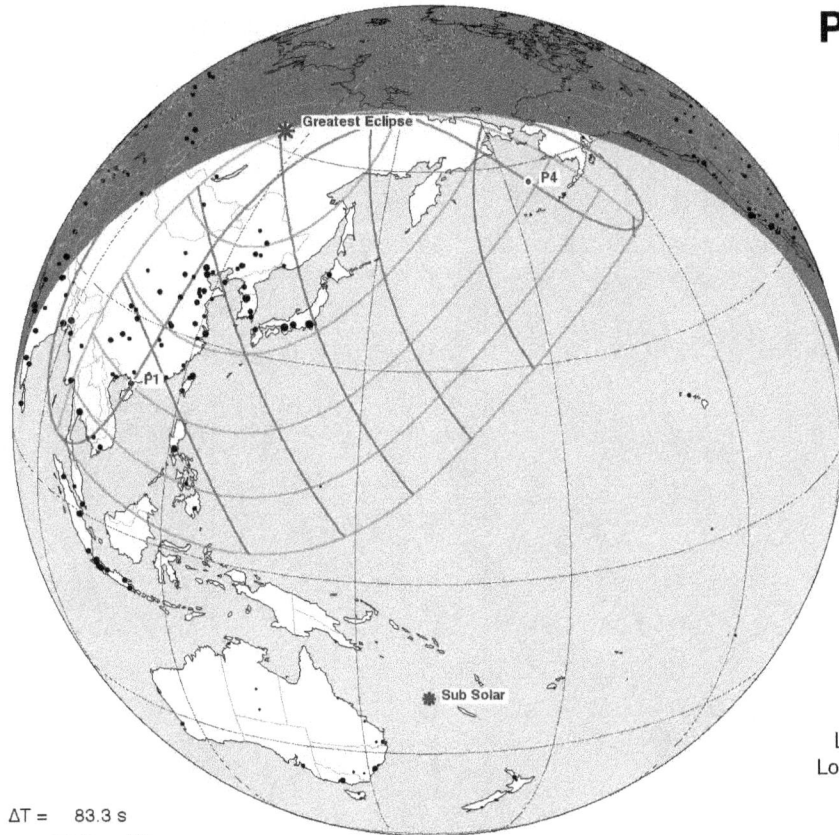

ΔT =   83.3 s

www.EclipseWise.com

©2020  F. Espenak

# Partial Solar Eclipse
## 2047 Jun 23

Greatest Eclipse = 10:51:07.1 UT1
Eclipse Magnitude = 0.3129
Eclipse Obscuration = 0.1978
Gamma = 1.3766
Saros = 118   [70 of 72]
Decending Node

### Geocentric Coordinates at Greatest Eclipse

| Sun | Moon |
|---|---|
| R.A. = 06h08m27.7s | R.A. = 06h09m05.2s |
| Dec. = +23°25'10.2" | Dec. = +24°40'56.6" |
| S.D. = 00°15'44.2" | S.D. = 00°15'07.9" |
| H.P. = 00°00'08.7" | H.P. = 00°55'32.1" |

### External/Internal Contacts of Penumbra

P1 = 09:28:08.5 UT1
P4 = 12:14:08.7 UT1

### Circumstances at Greatest Eclipse

| Lat. = 65°46.0'N | Sun Alt. =   0.0° |
|---|---|
| Long. = 178°02.2'W | Sun Azm. = 345.5° |

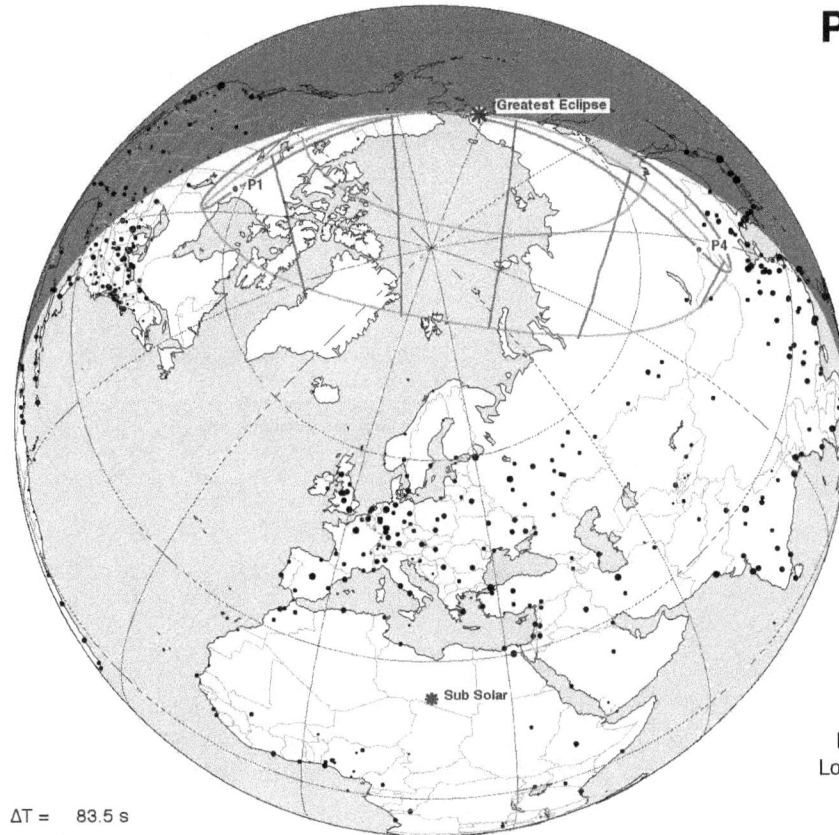

ΔT =   83.5 s

www.EclipseWise.com

©2020  F. Espenak

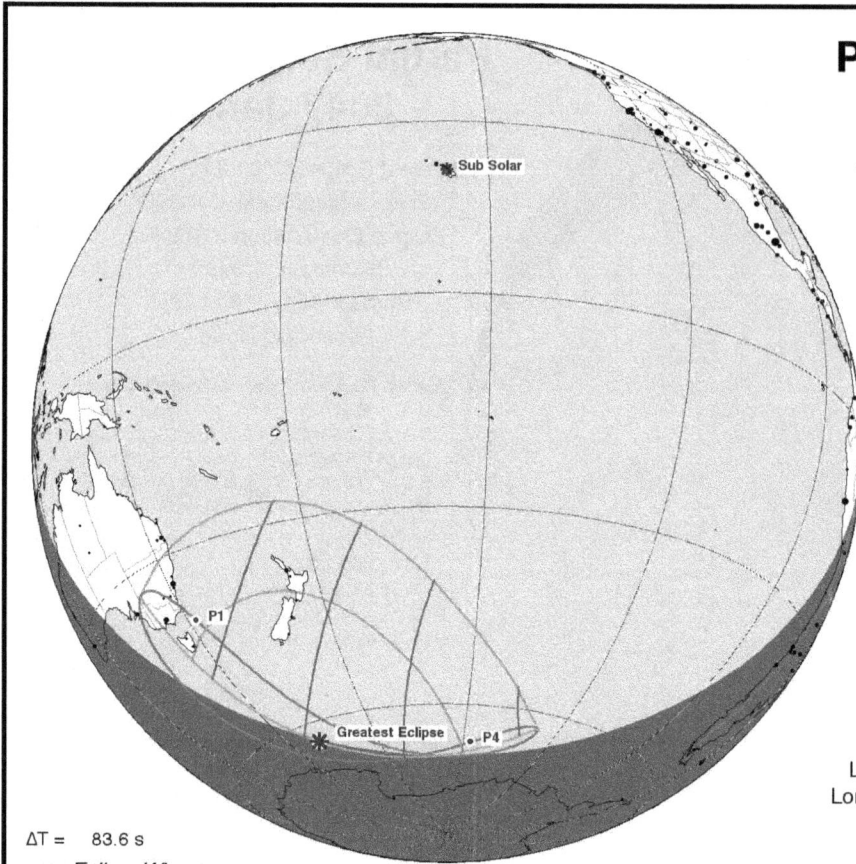

# Partial Solar Eclipse
## 2047 Jul 22

Greatest Eclipse = 22:34:53.8 UT1

Eclipse Magnitude = 0.3605

Eclipse Obscuration = 0.2443

Gamma = -1.3477

Saros = 156 [ 3 of 69]

Decending Node

Geocentric Coordinates at Greatest Eclipse

| | Sun | Moon |
|---|---|---|
| R.A. = | 08h08m59.7s | R.A. = 08h07m21.2s |
| Dec. = | +20°07'53.9" | Dec. = +18°54'51.1" |
| S.D. = | 00°15'44.5" | S.D. = 00°15'32.1" |
| H.P. = | 00°00'08.7" | H.P. = 00°57'00.9" |

External/Internal
Contacts of Penumbra

P1 = 21:09:54.0 UT1
P4 = 23:59:28.8 UT1

Circumstances at Greatest Eclipse

| Lat. = 63°26.5'S | Sun Alt. = 0.0° |
|---|---|
| Long. = 160°08.5'E | Sun Azm. = 39.7° |

ΔT = 83.6 s

www.EclipseWise.com

©2020 F. Espenak

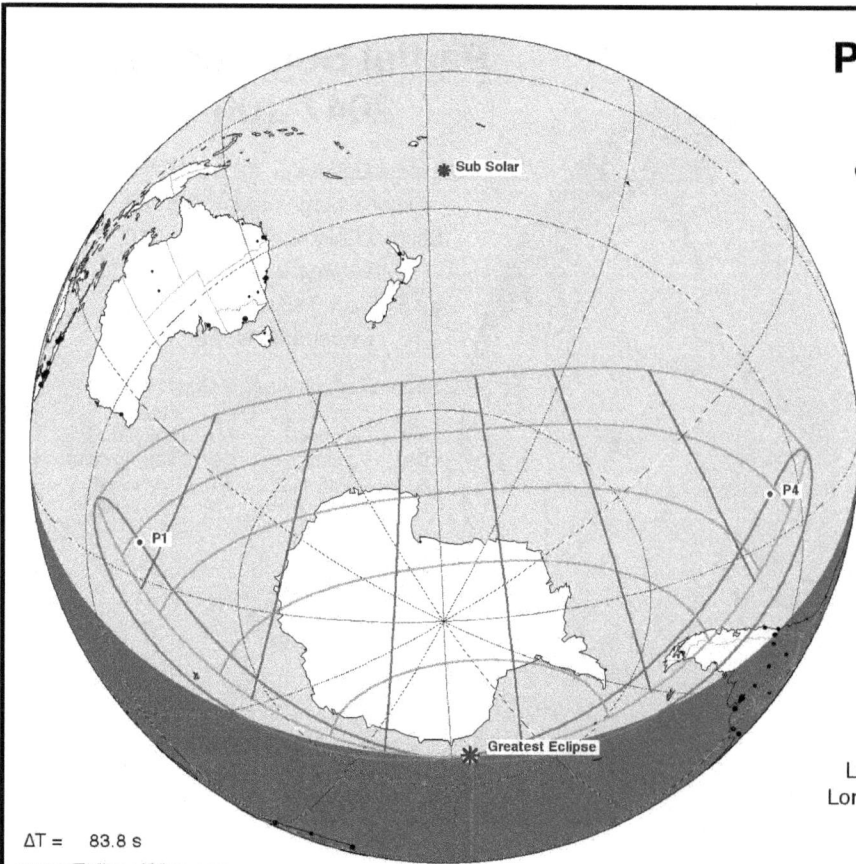

# Partial Solar Eclipse
## 2047 Dec 16

Greatest Eclipse = 23:48:48.5 UT1

Eclipse Magnitude = 0.8817

Eclipse Obscuration = 0.8553

Gamma = -1.0661

Saros = 123 [55 of 70]

Ascending Node

Geocentric Coordinates at Greatest Eclipse

| | Sun | Moon |
|---|---|---|
| R.A. = | 17h37m56.6s | R.A. = 17h38m13.1s |
| Dec. = | -23°20'10.9" | Dec. = -24°24'51.1" |
| S.D. = | 00°16'15.0" | S.D. = 00°16'35.9" |
| H.P. = | 00°00'08.9" | H.P. = 01°00'54.9" |

External/Internal
Contacts of Penumbra

P1 = 21:53:27.3 UT1
P4 = 01:44:15.0 UT1

Circumstances at Greatest Eclipse

| Lat. = 66°26.6'S | Sun Alt. = 0.0° |
|---|---|
| Long. = 006°37.9'W | Sun Azm. = 187.7° |

ΔT = 83.8 s

www.EclipseWise.com

©2020 F. Espenak

# Annular Solar Eclipse
## 2048 Jun 11

Greatest Eclipse = 12:57:28.6 UT1
Eclipse Magnitude = 0.9441
Eclipse Obscuration = 0.8914
Gamma = 0.6468
Saros = 128 [60 of 73]
Decending Node

### Geocentric Coordinates at Greatest Eclipse

| Sun | Moon |
|---|---|
| R.A. = 05h22m03.9s | R.A. = 05h22m09.1s |
| Dec. = +23°08'47.0" | Dec. = +23°43'34.6" |
| S.D. = 00°15'45.1" | S.D. = 00°14'42.3" |
| H.P. = 00°00'08.7" | H.P. = 00°53'58.0" |

| External/Internal Contacts of Penumbra | External/Internal Contacts of Umbra |
|---|---|
| P1 = 10:08:20.7 UT1 | U1 = 11:24:09.0 UT1 |
| P4 = 15:46:36.8 UT1 | U2 = 11:30:17.0 UT1 |
| | U3 = 14:24:42.5 UT1 |
| | U4 = 14:30:49.5 UT1 |

### Circumstances at Greatest Eclipse

| | |
|---|---|
| Lat. = 63°41.3'N | Sun Alt. = 49.4° |
| Long. = 011°32.3'W | Sun Azm. = 184.0° |
| Path Width = 271.5 km | Duration = 04m58.3s |

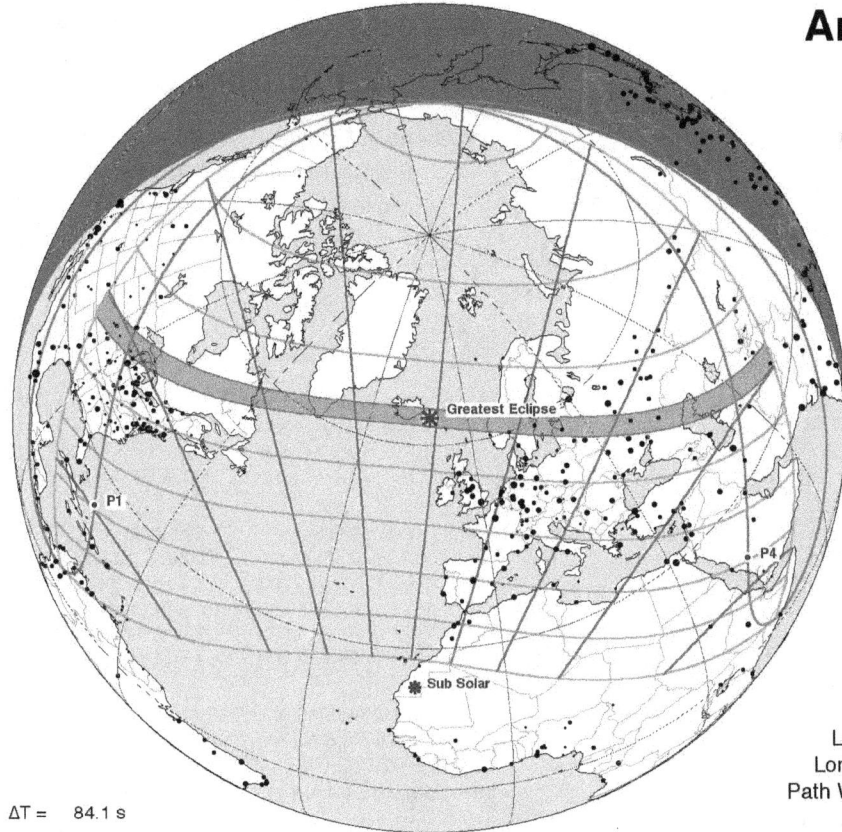

ΔT = 84.1 s

www.EclipseWise.com

©2020 F. Espenak

# Total Solar Eclipse
## 2048 Dec 05

Greatest Eclipse = 15:34:02.3 UT1
Eclipse Magnitude = 1.0440
Eclipse Obscuration = 1.0899
Gamma = -0.3973
Saros = 133 [47 of 72]
Ascending Node

### Geocentric Coordinates at Greatest Eclipse

| Sun | Moon |
|---|---|
| R.A. = 16h51m20.5s | R.A. = 16h51m18.6s |
| Dec. = -22°29'40.9" | Dec. = -22°53'56.4" |
| S.D. = 00°16'13.8" | S.D. = 00°16'40.9" |
| H.P. = 00°00'08.9" | H.P. = 01°01'13.3" |

| External/Internal Contacts of Penumbra | External/Internal Contacts of Umbra |
|---|---|
| P1 = 13:00:23.5 UT1 | U1 = 13:58:26.4 UT1 |
| P2 = 15:10:17.5 UT1 | U2 = 14:00:03.2 UT1 |
| P3 = 15:57:46.3 UT1 | U3 = 17:07:59.8 UT1 |
| P4 = 18:07:38.9 UT1 | U4 = 17:09:38.3 UT1 |

### Circumstances at Greatest Eclipse

| | |
|---|---|
| Lat. = 46°07.9'S | Sun Alt. = 66.4° |
| Long. = 056°24.1'W | Sun Azm. = 1.4° |
| Path Width = 160.3 km | Duration = 03m27.6s |

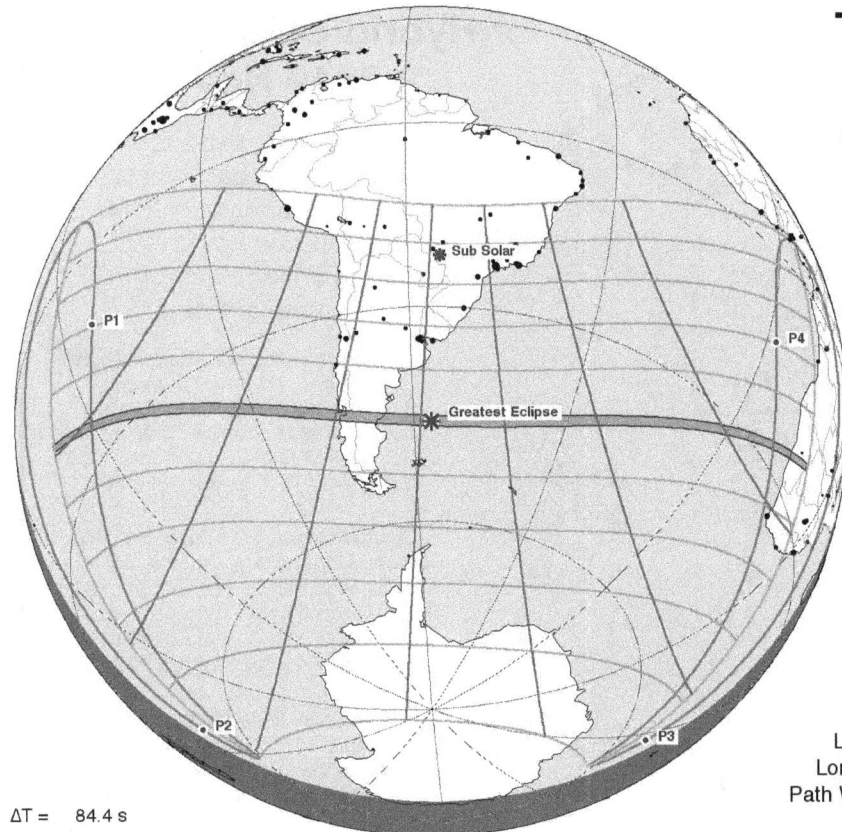

ΔT = 84.4 s

www.EclipseWise.com

©2020 F. Espenak

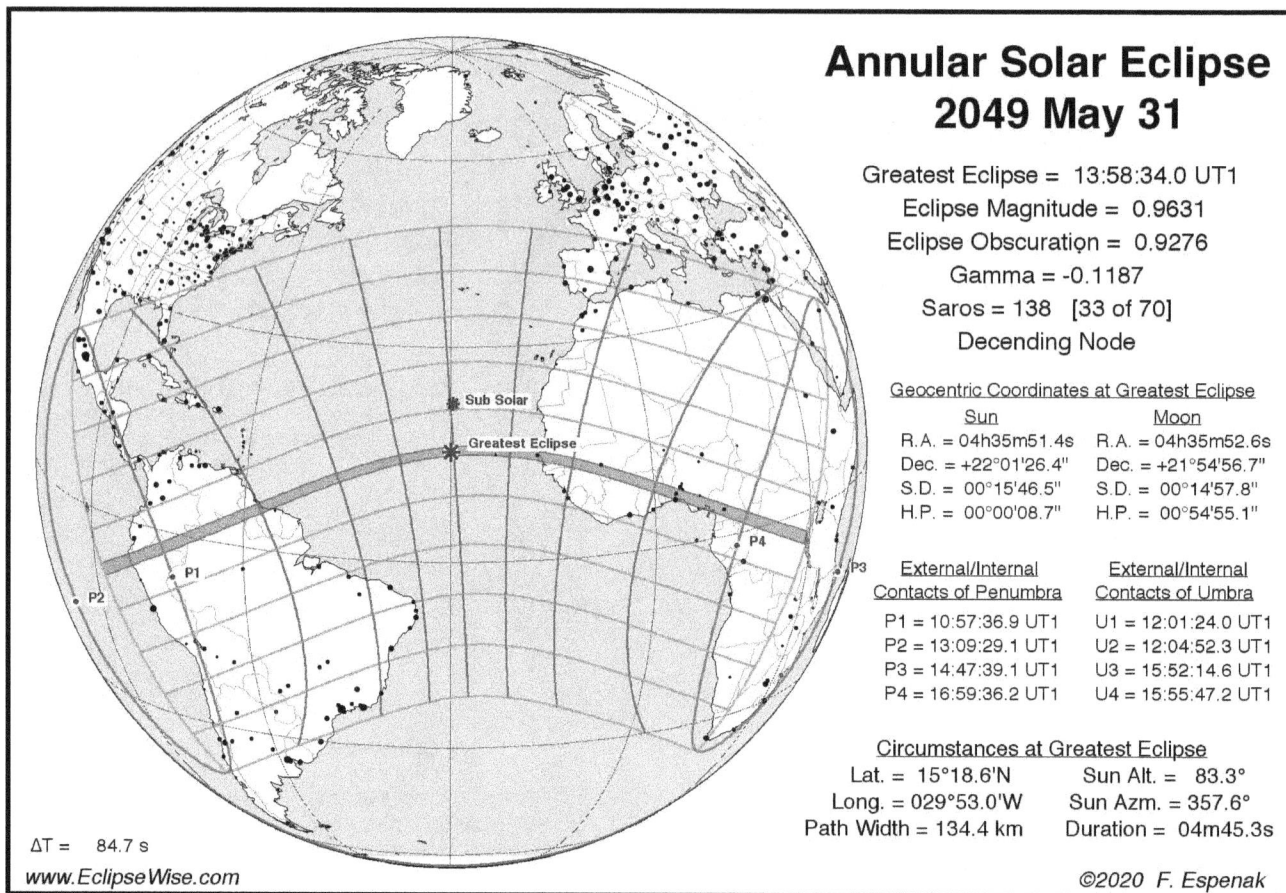

# Annular Solar Eclipse
# 2049 May 31

Greatest Eclipse = 13:58:34.0 UT1
Eclipse Magnitude = 0.9631
Eclipse Obscuration = 0.9276
Gamma = -0.1187
Saros = 138 [33 of 70]
Decending Node

### Geocentric Coordinates at Greatest Eclipse

| Sun | Moon |
|---|---|
| R.A. = 04h35m51.4s | R.A. = 04h35m52.6s |
| Dec. = +22°01'26.4" | Dec. = +21°54'56.7" |
| S.D. = 00°15'46.5" | S.D. = 00°14'57.8" |
| H.P. = 00°00'08.7" | H.P. = 00°54'55.1" |

| External/Internal Contacts of Penumbra | External/Internal Contacts of Umbra |
|---|---|
| P1 = 10:57:36.9 UT1 | U1 = 12:01:24.0 UT1 |
| P2 = 13:09:29.1 UT1 | U2 = 12:04:52.3 UT1 |
| P3 = 14:47:39.1 UT1 | U3 = 15:52:14.6 UT1 |
| P4 = 16:59:36.2 UT1 | U4 = 15:55:47.2 UT1 |

### Circumstances at Greatest Eclipse

| | |
|---|---|
| Lat. = 15°18.6'N | Sun Alt. = 83.3° |
| Long. = 029°53.0'W | Sun Azm. = 357.6° |
| Path Width = 134.4 km | Duration = 04m45.3s |

ΔT = 84.7 s

www.EclipseWise.com

©2020 F. Espenak

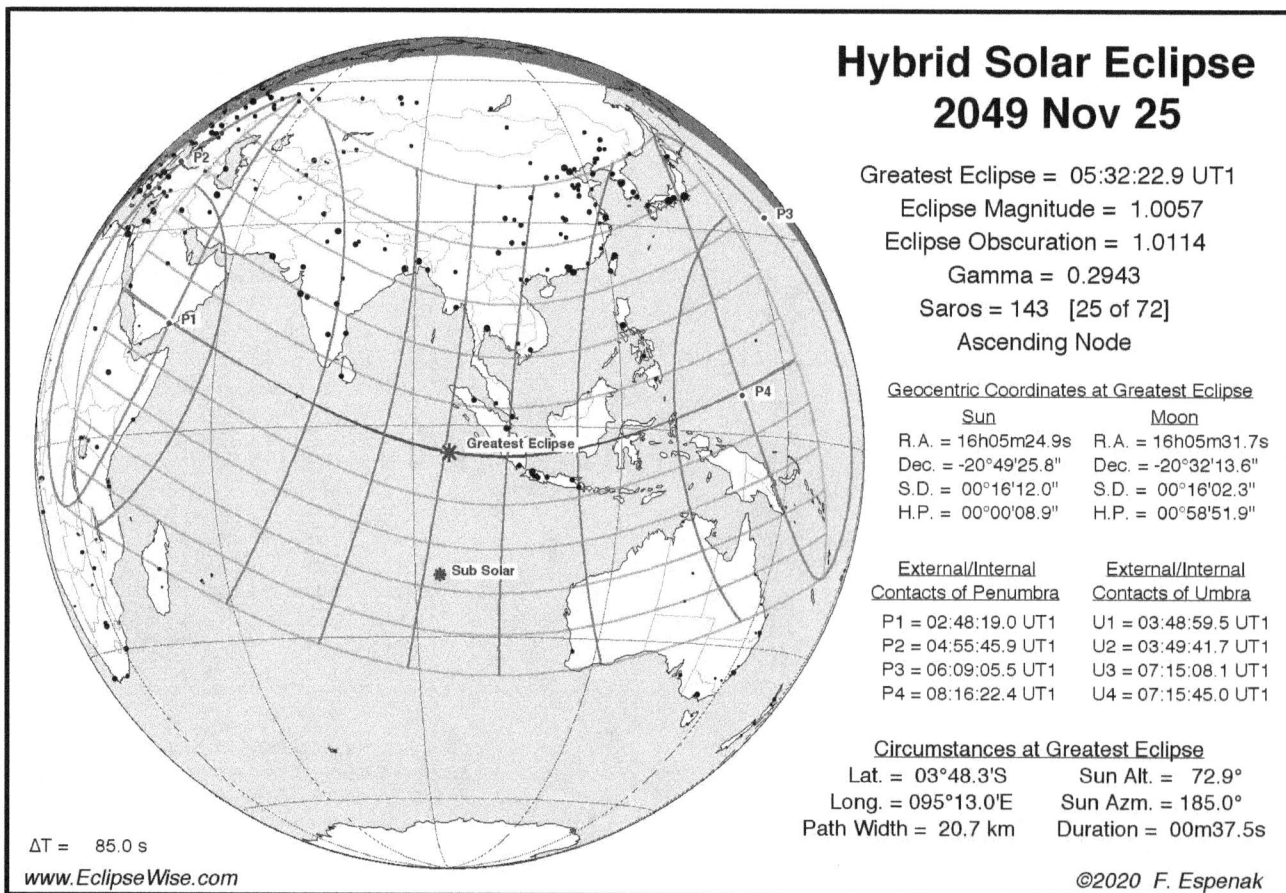

# Hybrid Solar Eclipse
# 2049 Nov 25

Greatest Eclipse = 05:32:22.9 UT1
Eclipse Magnitude = 1.0057
Eclipse Obscuration = 1.0114
Gamma = 0.2943
Saros = 143 [25 of 72]
Ascending Node

### Geocentric Coordinates at Greatest Eclipse

| Sun | Moon |
|---|---|
| R.A. = 16h05m24.9s | R.A. = 16h05m31.7s |
| Dec. = -20°49'25.8" | Dec. = -20°32'13.6" |
| S.D. = 00°16'12.0" | S.D. = 00°16'02.3" |
| H.P. = 00°00'08.9" | H.P. = 00°58'51.9" |

| External/Internal Contacts of Penumbra | External/Internal Contacts of Umbra |
|---|---|
| P1 = 02:48:19.0 UT1 | U1 = 03:48:59.5 UT1 |
| P2 = 04:55:45.9 UT1 | U2 = 03:49:41.7 UT1 |
| P3 = 06:09:05.5 UT1 | U3 = 07:15:08.1 UT1 |
| P4 = 08:16:22.4 UT1 | U4 = 07:15:45.0 UT1 |

### Circumstances at Greatest Eclipse

| | |
|---|---|
| Lat. = 03°48.3'S | Sun Alt. = 72.9° |
| Long. = 095°13.0'E | Sun Azm. = 185.0° |
| Path Width = 20.7 km | Duration = 00m37.5s |

ΔT = 85.0 s

www.EclipseWise.com

©2020 F. Espenak

# Hybrid Solar Eclipse
# 2050 May 20

Greatest Eclipse = 20:41:24.9 UT1

Eclipse Magnitude = 1.0038

Eclipse Obscuration = 1.0076

Gamma = -0.8688

Saros = 148  [23 of 75]

Decending Node

### Geocentric Coordinates at Greatest Eclipse

| | Sun | Moon |
|---|---|---|
| R.A. = | 03h51m25.4s | 03h51m49.6s |
| Dec. = | +20°09'01.9" | +19°19'17.1" |
| S.D. = | 00°15'48.3" | 00°15'44.7" |
| H.P. = | 00°00'08.7" | 00°57'47.0" |

| External/Internal Contacts of Penumbra | External/Internal Contacts of Umbra |
|---|---|
| P1 = 18:21:05.6 UT1 | U1 = 19:47:22.5 UT1 |
| P4 = 23:01:54.7 UT1 | U2 = 19:47:52.6 UT1 |
| | U3 = 21:35:02.3 UT1 |
| | U4 = 21:35:38.0 UT1 |

### Circumstances at Greatest Eclipse

| | |
|---|---|
| Lat. = 40°05.5'S | Sun Alt. = 29.4° |
| Long. = 123°44.9'W | Sun Azm. = 352.0° |
| Path Width = 26.7 km | Duration = 00m21.3s |

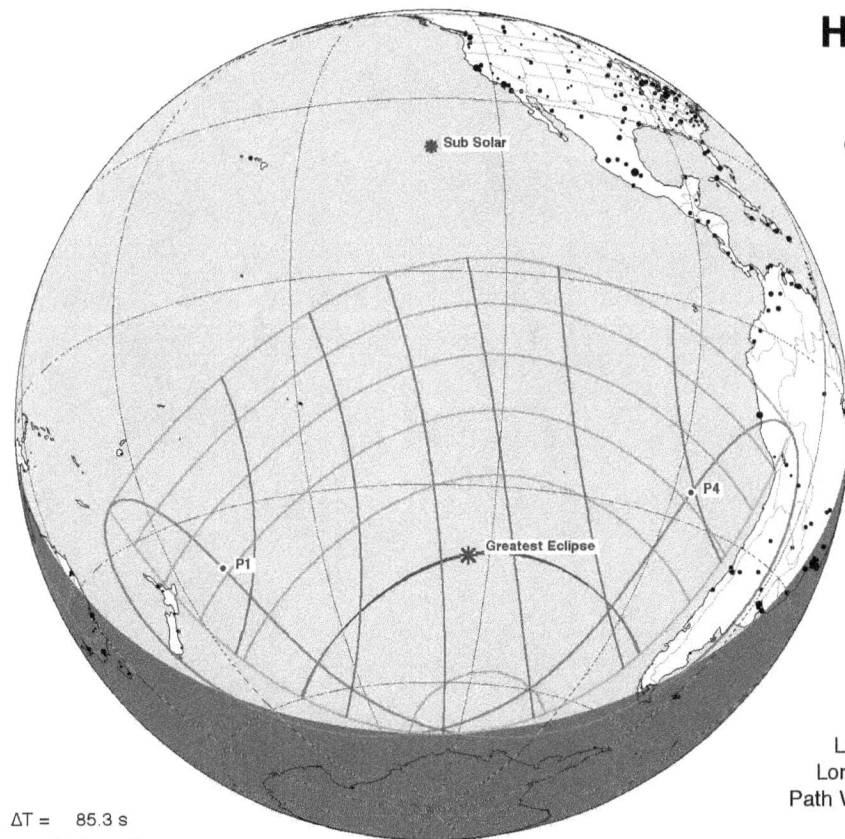

ΔT = 85.3 s

www.EclipseWise.com

©2020  F. Espenak

# Partial Solar Eclipse
# 2050 Nov 14

Greatest Eclipse = 13:29:27.1 UT1

Eclipse Magnitude = 0.8874

Eclipse Obscuration = 0.8322

Gamma = 1.0447

Saros = 153  [11 of 70]

Ascending Node

### Geocentric Coordinates at Greatest Eclipse

| | Sun | Moon |
|---|---|---|
| R.A. = | 15h19m50.5s | 15h20m29.5s |
| Dec. = | -18°21'19.2" | -17°24'01.9" |
| S.D. = | 00°16'09.8" | 00°15'10.6" |
| H.P. = | 00°00'08.9" | 00°55'41.9" |

| External/Internal Contacts of Penumbra |
|---|
| P1 = 11:16:02.1 UT1 |
| P4 = 15:42:55.8 UT1 |

### Circumstances at Greatest Eclipse

| | |
|---|---|
| Lat. = 69°31.7'N | Sun Alt. = 0.0° |
| Long. = 000°59.3'E | Sun Azm. = 205.8° |

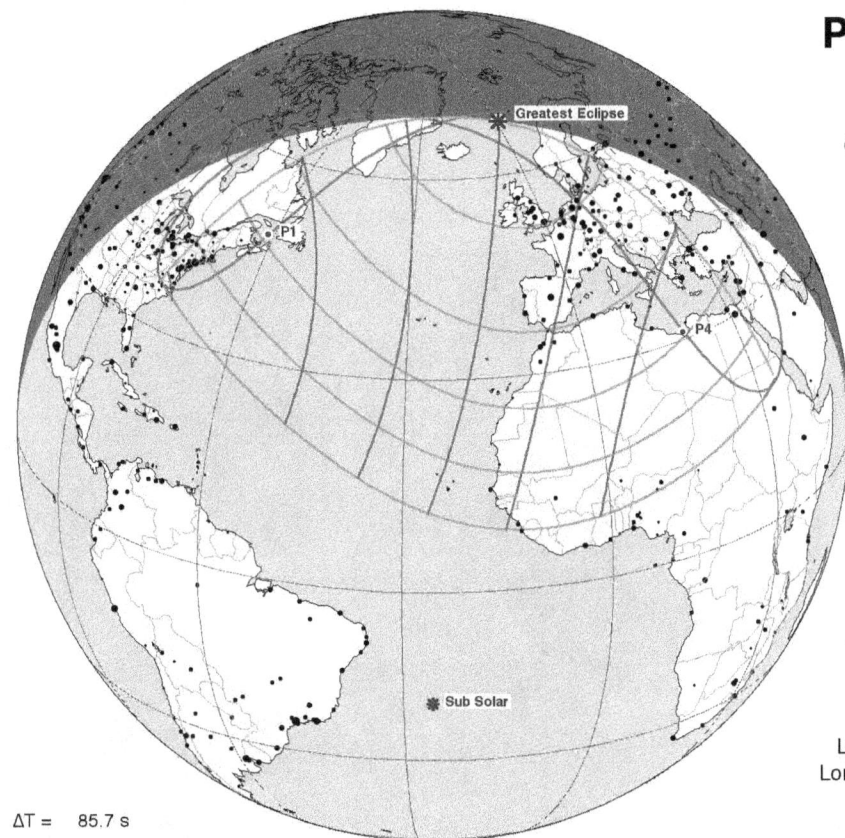

ΔT = 85.7 s

www.EclipseWise.com

©2020  F. Espenak

# Key to Catalog of Solar Eclipses

The following catalog lists all solar eclipses for a 25-year period.

A brief description of each parameter in the catalog appears below.

**Date** — Gregorian date of Greatest Eclipse

**Greatest Eclipse** — Universal Time of Greatest Eclipse

**Saros** — Saros Series Number of the eclipse

**Type** — Solar Eclipse Type

        P = Partial Solar Eclipse
        A = Annular Solar Eclipse
        T = Total Solar Eclipse
        H = Hybrid Solar Eclipse

        + = Non-central eclipse of with no northern limit
        – = Non-central eclipse of with no southern limit

        b = Saros series begins (first eclipse in a Saros series)
        e = Saros series ends (last eclipse in a Saros series)

**Gamma** — minimum distance from the axis of the lunar shadow to the center of Earth

**Mag** — Eclipse Magnitude; fraction of the Sun's diameter obscured by the Moon

**Lat & Long** — latitude and longitude where the Sun appears in the zenith at greatest eclipse

**Alt** — altitude of the Sun at greatest eclipse

**Duration** — Central Line Duration (minutes. seconds) at greatest eclipse

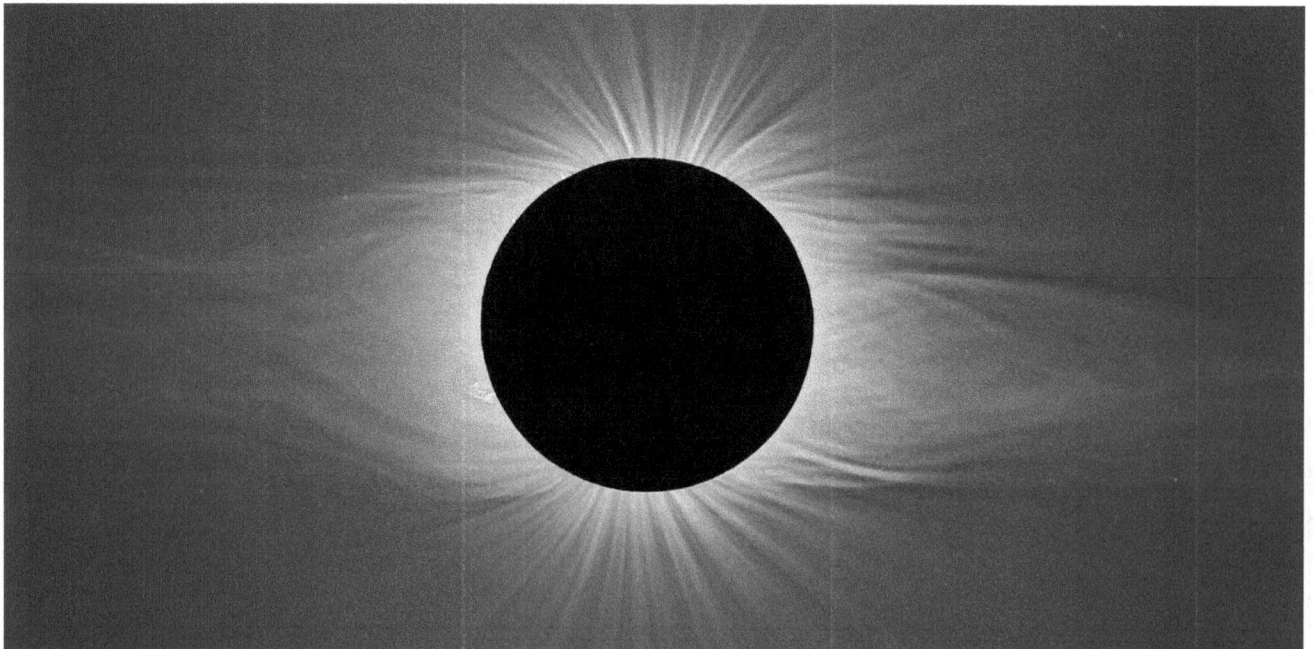

*Photo 1–8 The solar corona is revealed during totality. Total Solar Eclipse of 2018 Jul 03. ©2018 F. Espenak*

# Catalog of Solar Eclipses: 2041 to 2065

| Date | Greatest Eclipse | Saros | Type | Gamma | Mag | Lat | Long | Alt | Duration |
|------|------------------|-------|------|-------|-----|-----|------|-----|----------|
| 2041 Apr 30 | 11:51:01 | 129 | T | -0.4492 | 1.0189 | 9.6S | 12.2E | 63 | 01m51s |
| 2041 Oct 25 | 1:35:02 | 134 | A | 0.4133 | 0.9467 | 9.9N | 162.8E | 66 | 06m07s |
| 2042 Apr 20 | 2:16:10 | 139 | T | 0.2956 | 1.0614 | 27.0N | 137.3E | 73 | 04m51s |
| 2042 Oct 14 | 1:59:21 | 144 | A | -0.3030 | 0.9301 | 23.7S | 137.8E | 72 | 07m44s |
| 2043 Apr 09 | 18:56:28 | 149 | T+ | 1.0031 | 1.0096 | 61.3N | 151.9E | 0 | 01m46s |
| 2043 Oct 03 | 3:00:28 | 154 | A- | -1.0102 | 0.9497 | 61.0S | 35.2E | 0 | 02m17s |
| 2044 Feb 28 | 20:23:18 | 121 | As | -0.9954 | 0.9600 | 62.2S | 25.6W | 4 | 02m27s |
| 2044 Aug 23 | 1:15:40 | 126 | T | 0.9613 | 1.0364 | 64.3N | 120.5W | 15 | 02m04s |
| 2045 Feb 16 | 23:54:45 | 131 | A | -0.3125 | 0.9285 | 28.3S | 166.2W | 72 | 07m47s |
| 2045 Aug 12 | 17:41:17 | 136 | T | 0.2116 | 1.0774 | 25.9N | 78.6W | 78 | 06m06s |
| 2046 Feb 05 | 23:05:03 | 141 | A | 0.3765 | 0.9232 | 4.8N | 171.4W | 68 | 09m42s |
| 2046 Aug 02 | 10:19:50 | 146 | T | -0.5350 | 1.0531 | 12.7S | 15.1E | 58 | 04m51s |
| 2047 Jan 26 | 1:31:55 | 151 | P | 1.0450 | 0.8908 | 62.9N | 111.7E | 0 | |
| 2047 Jun 23 | 10:51:07 | 118 | P | 1.3766 | 0.3129 | 65.8N | 178.0W | 0 | |
| 2047 Jul 22 | 22:34:54 | 156 | P | -1.3477 | 0.3605 | 63.4S | 160.1E | 0 | |
| 2047 Dec 16 | 23:48:48 | 123 | P | -1.0661 | 0.8817 | 66.4S | 6.6W | 0 | |
| 2048 Jun 11 | 12:57:29 | 128 | A | 0.6468 | 0.9441 | 63.7N | 11.5W | 49 | 04m58s |
| 2048 Dec 05 | 15:34:02 | 133 | T | -0.3973 | 1.0440 | 46.1S | 56.4W | 66 | 03m28s |
| 2049 May 31 | 13:58:34 | 138 | A | -0.1187 | 0.9631 | 15.3N | 29.9W | 83 | 04m45s |
| 2049 Nov 25 | 5:32:23 | 143 | H | 0.2943 | 1.0057 | 3.8S | 95.2E | 73 | 00m38s |
| 2050 May 20 | 20:41:25 | 148 | H | -0.8688 | 1.0038 | 40.1S | 123.7W | 29 | 00m21s |
| 2050 Nov 14 | 13:29:27 | 153 | P | 1.0447 | 0.8874 | 69.5N | 1.0E | 0 | |
| 2051 Apr 11 | 2:09:13 | 120 | P | 1.0169 | 0.9849 | 71.6N | 32.1E | 0 | |
| 2051 Oct 04 | 21:00:48 | 125 | P | -1.2094 | 0.6024 | 72.0S | 117.7E | 0 | |
| 2052 Mar 30 | 18:30:26 | 130 | T | 0.3239 | 1.0466 | 22.4N | 102.6W | 71 | 04m08s |
| 2052 Sep 22 | 23:37:43 | 135 | A | -0.4480 | 0.9734 | 25.7S | 174.9E | 63 | 02m51s |
| 2053 Mar 20 | 7:06:52 | 140 | A | -0.4089 | 0.9919 | 23.0S | 82.9E | 66 | 00m50s |
| 2053 Sep 12 | 9:32:41 | 145 | T | 0.3140 | 1.0328 | 21.5N | 41.7E | 72 | 03m04s |
| 2054 Mar 09 | 12:32:13 | 150 | P | -1.1711 | 0.6678 | 72.0S | 97.9E | 0 | |
| 2054 Aug 03 | 18:02:34 | 117 | Pe | -1.4941 | 0.0656 | 69.8S | 121.4W | 0 | |
| 2054 Sep 02 | 1:08:05 | 155 | P | 1.0215 | 0.9793 | 71.7N | 82.4W | 0 | |
| 2055 Jan 27 | 17:52:37 | 122 | P | 1.1550 | 0.6932 | 69.5N | 112.3W | 0 | |
| 2055 Jul 24 | 9:56:21 | 127 | T | -0.8012 | 1.0359 | 33.3S | 25.7E | 37 | 03m17s |
| 2056 Jan 16 | 22:15:16 | 132 | A | 0.4199 | 0.9760 | 3.9N | 153.6W | 65 | 02m52s |
| 2056 Jul 12 | 20:20:30 | 137 | A | -0.0426 | 0.9878 | 19.4N | 123.8W | 88 | 01m26s |
| 2057 Jan 05 | 9:46:22 | 142 | T | -0.2837 | 1.0287 | 39.2S | 35.1E | 73 | 02m29s |
| 2057 Jul 01 | 23:38:45 | 147 | A | 0.7455 | 0.9464 | 71.5N | 176.3W | 41 | 04m22s |
| 2057 Dec 26 | 1:13:05 | 152 | T | -0.9405 | 1.0348 | 84.9S | 21.7E | 19 | 01m50s |
| 2058 May 22 | 10:37:55 | 119 | P | -1.3194 | 0.4141 | 63.5S | 61.1E | 0 | |
| 2058 Jun 21 | 0:18:04 | 157 | Pb | 1.4869 | 0.1261 | 65.9N | 9.8E | 0 | |
| 2058 Nov 16 | 3:21:36 | 124 | P | 1.1224 | 0.7644 | 62.9N | 174.1E | 0 | |
| 2059 May 11 | 19:20:44 | 129 | T | -0.5080 | 1.0242 | 10.7S | 100.5W | 59 | 02m23s |
| 2059 Nov 05 | 9:16:43 | 134 | A | 0.4454 | 0.9417 | 8.7N | 47.0E | 63 | 07m00s |
| 2060 Apr 30 | 10:08:28 | 139 | T | 0.2422 | 1.0660 | 28.0N | 20.8E | 76 | 05m15s |
| 2060 Oct 24 | 9:22:38 | 144 | A | -0.2625 | 0.9277 | 25.8S | 28.0E | 75 | 08m06s |
| 2061 Apr 20 | 2:55:16 | 149 | T | 0.9578 | 1.0476 | 64.5N | 59.1E | 16 | 02m37s |
| 2061 Oct 13 | 10:30:36 | 154 | A | -0.9639 | 0.9469 | 62.1S | 54.5W | 15 | 03m41s |
| 2062 Mar 11 | 4:24:43 | 121 | P | -1.0238 | 0.9331 | 61.0S | 147.2W | 0 | |
| 2062 Sep 03 | 8:52:54 | 126 | P | 1.0192 | 0.9749 | 61.3N | 150.2E | 0 | |
| 2063 Feb 28 | 7:41:56 | 131 | A | -0.3360 | 0.9293 | 25.2S | 77.6E | 70 | 07m41s |
| 2063 Aug 24 | 1:20:36 | 136 | T | 0.2771 | 1.0750 | 25.6N | 168.3E | 74 | 05m49s |
| 2064 Feb 17 | 6:58:48 | 141 | A | 0.3596 | 0.9262 | 7.0N | 69.6E | 69 | 08m56s |
| 2064 Aug 12 | 17:44:31 | 146 | T | -0.4652 | 1.0495 | 10.9S | 96.1W | 62 | 04m28s |
| 2065 Feb 05 | 9:50:50 | 151 | P | 1.0336 | 0.9123 | 62.2N | 22.0W | 0 | |
| 2065 Jul 03 | 17:32:17 | 118 | P | 1.4619 | 0.1639 | 64.8N | 71.7E | 0 | |
| 2065 Aug 02 | 5:32:41 | 156 | P | -1.2758 | 0.4903 | 62.7S | 46.4E | 0 | |
| 2065 Dec 27 | 8:38:19 | 123 | P | -1.0688 | 0.8769 | 65.4S | 149.3W | 0 | |

*Photo 2–1 Phases of the total lunar eclipse of 2014 Apr 15 are captured in this multiple exposure sequence. ©2014 F. Espenak*

# Section 2: Lunar Eclipses

## Introduction

The Moon orbits Earth once every 29.5306 days with respect to the Sun. Over the course of its orbit, the Moon's changing position relative to the Sun results in its familiar phases: New Moon > First Quarter > Full Moon > Last Quarter > New Moon. The New Moon phase is the only one not visible because the illuminated side of the Moon points away from Earth.

During Full Moon, the Moon appears opposite the Sun in the sky. It rises as the Sun sets and is visible throughout the night. The Full Moon sets in the morning just as the Sun rises. This geometry occurs when the Moon is 180° from the Sun as seen from Earth. It corresponds to the direction Earth casts its shadow into space.

The Moon's orbit is tilted about 5.1° to Earth's orbit around the Sun. As seen from Earth, the points where the two orbits appear to cross are called the nodes. When the Full Moon occurs near one of these nodes, the Moon can pass through some portion of Earth's shadow and a lunar eclipse occurs.

Earth's shadow has two cone-shaped components, one nested inside the other. The outer or penumbral shadow is a zone where the Sun's rays are partially blocked., The inner or umbral shadow is a region where direct rays from the Sun are completely blocked.

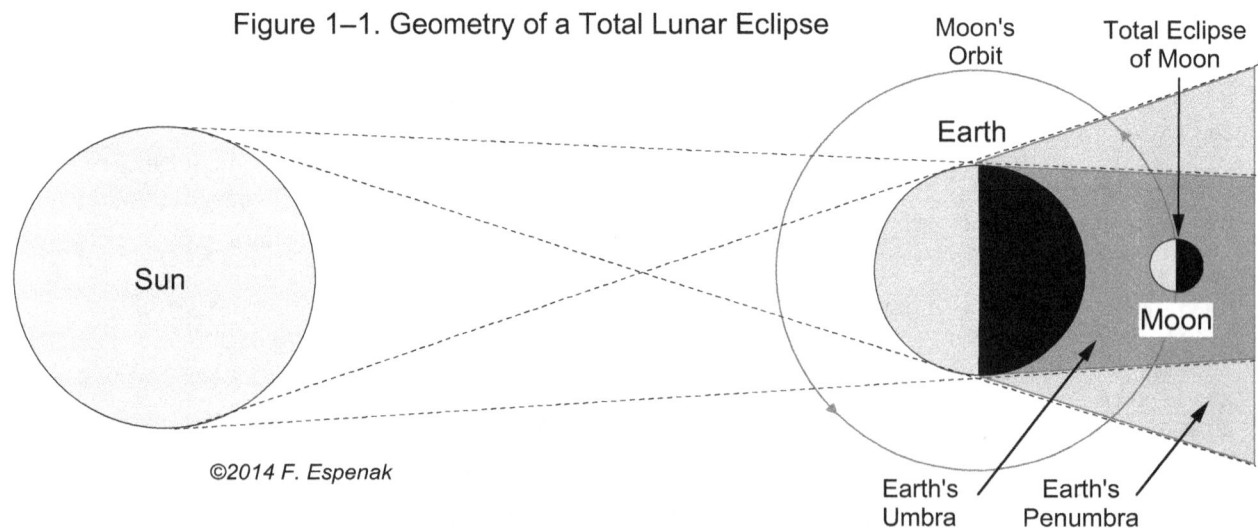

Figure 1–1. Geometry of a Total Lunar Eclipse

*Figure 2–1 illustrates the geometry of a total lunar eclipse. A partial eclipse is visible if only part of the Moon enters Earth's umbral shadow. If the Moon passes through the penumbral shadow but misses the umbral shadow, then a penumbral eclipse occurs.*

*Photo 2–2 Examples of the visual appearance of a penumbral, partial, and total lunar eclipse. ©2020 F. Espenak*

## Types of Lunar Eclipses

There are three types of lunar eclipses:

1. **Penumbral Lunar Eclipse** — The Moon passes through Earth's faint penumbral shadow. Penumbral eclipses are of minor interest since they are quite subtle and difficult to observe.
2. **Partial Lunar Eclipse** — A portion of the Moon passes through Earth's dark umbral shadow. The remaining part of the Moon appears bright even though it lies deep within the penumbra. Partial eclipses are easy to see, even with the unaided eye.
3. **Total Lunar Eclipse** — The entire Moon passes through Earth's umbral shadow. Total eclipses are quite striking for the vibrant range of color of the Moon during the total phase, referred to as totality.

## Figure 1–2. Types of Lunar Eclipses

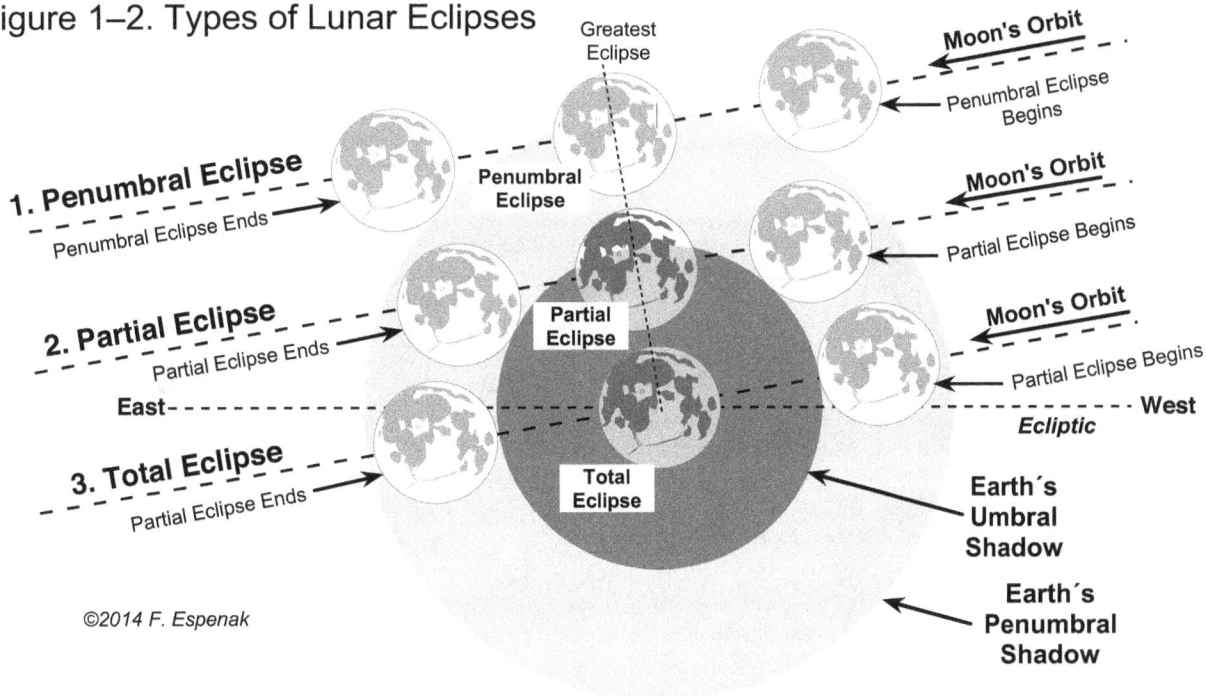

*Figure 2–2 illustrates the three types of lunar eclipses as seen from Earth. 1) A penumbral eclipse occurs when the Moon passes through the penumbra but completely misses the umbra. 2) A partial eclipse occurs if some portion of the Moon enters the umbra. 3) A total lunar eclipse takes place when the entire disk of the Moon enters the umbral shadow.*

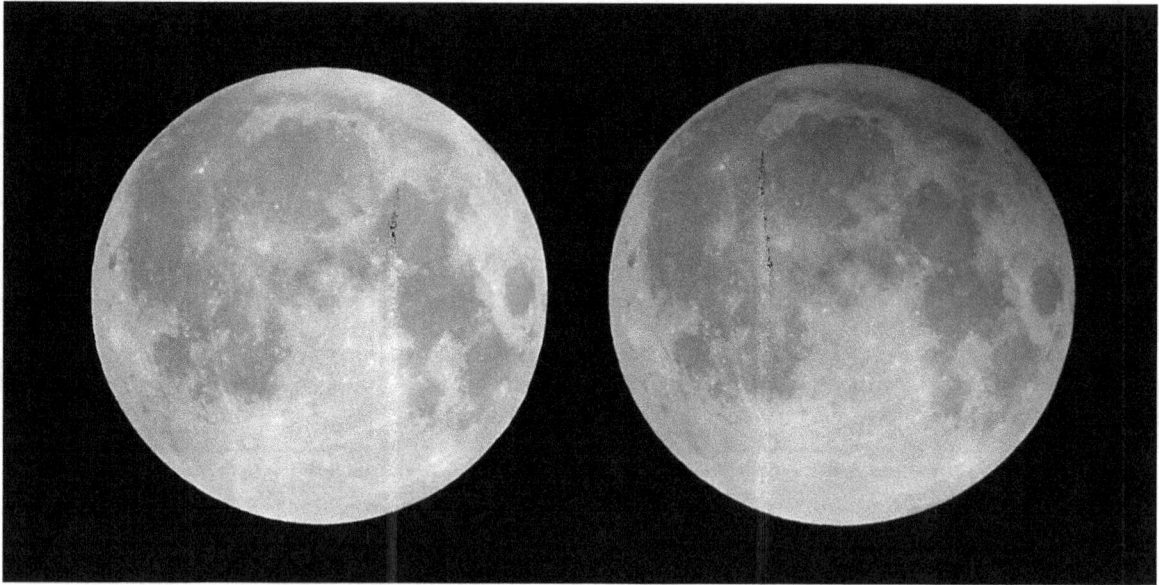

*Photo 2–3 Penumbral lunar eclipse of 2002 Nov 20. Left: Moon before the eclipse begins. Right: mid-eclipse when 88.9% of the Moon's is in the penumbral shadow. ©2002 F. Espenak*

## Visual Appearance of Penumbral Lunar Eclipses

The visual appearance of penumbral, partial and total lunar eclipses differs significantly. While penumbral eclipses are pale and difficult to see, partial eclipses are easy naked-eye events while total eclipses are colorful and dramatic.

Earth's penumbral shadow forms a diverging cone that expands into space away from the Sun. Within this zone, Earth blocks part but not all of the Sun's disk. Some portion of the Sun's direct rays continue to reach the Moon during a penumbral eclipse.

The early and late stages of a penumbral eclipse are completely invisible to the eye. It is only when about 2/3 of the Moon's disk has entered the penumbral shadow that a skilled observer can detect a faint shading across the Moon.

Even when 90% of the Moon is immersed in the penumbra, approximately 10% of the Sun's rays still reach the Moon's deepest limb. Under such conditions, the Moon remains relatively bright with only a subtle shadow gradient across its disk.

*Photo 2–4 Time sequence of the partial lunar eclipse of 2012 Jun 04. ©2012 F. Espenak*

## Visual Appearance of Partial Lunar Eclipses

Compared to penumbral eclipses, partial eclipses are easy to see with the naked eye. The lunar limb extending into the umbral shadow appears very dark or even black. This is due to a contrast effect since the remaining portion of the Moon in the penumbra is hundreds of times brighter. Because the umbral shadow's diameter is about 2.7 times the Moon's diameter, it appears as though a semi-circular bite has been taken out of the Moon.

Aristotle (384–322 BCE) first proved that Earth was round using the curved umbral shadow seen at partial eclipses. In comparing observations of several eclipses, he noted that Earth's shadow was round no matter where the eclipse took place, whether the Moon was high in the sky or low near the horizon. Aristotle reasoned that only a sphere casts a round shadow from every angle.

*Photo 2–5 The beginning, middle and end of totality during the total lunar eclipse of 2004 October 28. ©2004 F. Espenak*

## Visual Appearance of Total Lunar Eclipses

A total lunar eclipse is the most dramatic and visually compelling type of lunar eclipse. The Moon's appearance can vary enormously throughout the period of totality and from one eclipse to the next. The geometry of the Moon's path through the umbra plays a significant role in determining the appearance of totality. The effect that Earth's atmosphere has on a total eclipse is not as apparent. Although the physical mass of Earth blocks all direct sunlight from the umbra, the planet's atmosphere filters, attenuates and bends some of the Sun's rays into the shadow.

The molecules in Earth's atmosphere scatter short wavelength light (i.e., yellow, green, blue) more than long wavelength light (i.e., orange, red). The same process responsible for making sunsets red also gives total lunar eclipses their characteristic ruddy color. The exact appearance can vary widely in both hue and brightness.

Because the lowest layers of the atmosphere are the thickest, they absorb more sunlight and refract it through larger angles. About 75% of the atmosphere's mass is concentrated in the bottom 10 kilometers (troposphere) as well as most of the water vapor, which can form massive clouds that block even more light. Just above the troposphere lies the stratosphere (10 to 50 kilometers), a rarified zone above most of the planet's weather systems. The stratosphere is subject to important photochemical reactions due to the high level of solar ultraviolet radiation that penetrates the region. The troposphere and stratosphere act together as a ring-shaped lens that refracts heavily reddened sunlight into Earth's umbral shadow. Since the higher stratospheric layers contain less gas, they refract sunlight through progressively smaller angles into the outer parts of the umbra. In contrast, denser tropospheric layers refract sunlight through larger angles to reach the inner parts of the umbra.

Because of this lensing effect, the amount of light refracted into the umbra tends to decrease radially from the edge to the center. Inhomogeneities from varying amounts of cloud and dust at differing latitudes can cause significant variations in brightness throughout the umbra.

Besides water (cloud, mist, precipitation), Earth's atmosphere also contains aerosols or tiny particles of organic debris, meteoric dust, volcanic ash and photochemical droplets. This material attenuates sunlight before it is refracted into the umbra. For instance, major volcanic eruptions in 1963 (Agung) and 1982 (El Chichon) each dumped large quantities of gas and ash into the stratosphere and were followed by several years of dark eclipses.

The 1991 eruption of Pinatubo in the Philippines had a similar effect. While most of the solid ash fell to Earth several days after circulating through the troposphere, a sizable volume of sulfur dioxide ($SO_2$) reached the stratosphere where it interacts with water vapor and eventually produces sulfuric acid ($H_2SO_4$). This high-altitude volcanic haze layer severely attenuates sunlight that must travel several hundred kilometers horizontally through the layer before being refracted into the umbral shadow. Thus, total eclipses following large volcanic eruptions are particularly dark. The total lunar eclipse of 1992 Dec 09 (1½ years after Pinatubo) was so dark that it was difficult to see the Moon's dull gray disk with the naked eye.

All total eclipses begin with penumbral and partial phases. After the total phase, the eclipse ends with more partial and penumbral phases. Lunar eclipses are completely safe to view and require none of the precautions needed for viewing solar eclipses (like special filters). The best views of a lunar eclipse are with binoculars and the naked eye.

## Figure 2–1. Lunar Eclipse Contacts

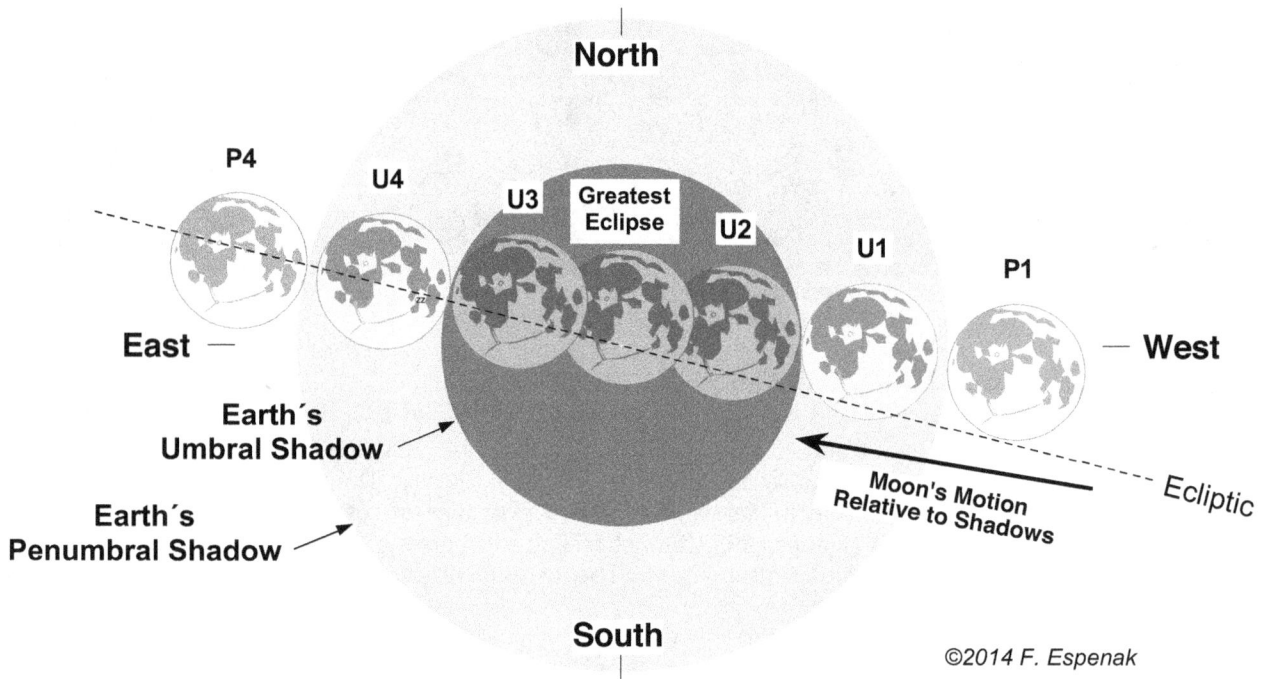

Figure 2–3 illustrates the six contacts for a total lunar eclipse. These correspond to the instants when the Moon's disk is externally tangent to the penumbra (P1 and P4), or either externally or internally tangent to the umbra (U1, U2, U3, and U4). Partial eclipses do not have contacts U2 and U3, while penumbral eclipses only have contacts P1 and P4.

## Lunar Eclipse Contacts

During the course of a lunar eclipse, the instants when the Moon's disk becomes tangent to Earth's shadows are known as eclipse contacts. They mark the primary stages or phases of a lunar eclipse (see figure 2-3) and are defined as follows.

| | |
|---|---|
| **P1** — Penumbral Eclipse Begins | (Instant of first exterior tangency of the Moon with the Penumbra) |
| **U1** — Partial Eclipse Begins | (Instant of first exterior tangency of the Moon with the Umbra) |
| **U2** — Total Eclipse Begins | (Instant of first interior tangency of the Moon with the Umbra) |
| **U3** — Total Eclipse Ends | (Instant of last interior tangency of the Moon with the Umbra) |
| **U4** — Partial Eclipse Ends | (Instant of last exterior tangency of the Moon with the Umbra) |
| **P4** — Penumbral Eclipse Ends | (Instant of last exterior tangency of the Moon with the Penumbra) |

Penumbral eclipses only have first and last contacts (i.e., P1 and P4) although neither of these events is observable.

In addition to P1 and P4, partial eclipses also have contacts U1 and U4 when the partial phases begin and end.

Total lunar eclipses have all six contacts. Contacts U2 and U3 mark the instants when the Moon's entire disk is first and last internally tangent to the umbra. These are the times when the total phase of the eclipse begins and ends..

The instant of greatest eclipse occurs when the Moon passes closest to the shadow axis. This is the maximum phase of the eclipse when the Moon is at its deepest position within either the penumbral or umbral shadow.

*Photo 2–6 Five minutes before the start of totality (2018 Jan 31), the Moon is bathed in an orange-red light.*
*The narrow rim outside the umbra and still in sunlight appears brilliant white. ©2018 F. Espenak*

## Enlargement of Earth's Shadows

In 1707, Philippe de La Hire made a curious observation about Earth's umbra. The predicted radius of the shadow needed to be enlarged by about 1/41 in order to fit timings made during a lunar eclipse. The enlargement is attributed to Earth's atmosphere, which becomes increasingly less transparent at lower levels. Additional observations over the next two centuries revealed that the shadow enlargement was somewhat variable from one eclipse to the next.

William Chauvenet (1891) formulated one method to account for the shadow enlargement while André-Louis Danjon (1951) devised another. Both methods assumed a circular cross section for the umbral shadow. However, Earth is flattened at the poles and bulges at the Equator, so an oblate spheroid more closely represents its shape. The projection of each of the planet's shadows is an ellipse rather than a circle. Furthermore, Earth's axial tilt towards or away from the Sun throughout the year means the elliptical shape of the penumbral and umbral shadows varies as well.

In an analysis of 22,539 observations made at 94 lunar eclipses from 1842 to 2011, Herald and Sinnott[7] (2014) found that the size and shape of the umbra are consistent with an oblate spheroid at the time of each eclipse, enlarged by the empirically determined occulting layer that uniformly surrounds Earth. The effective height of this layer was found to be 87 kilometers. Based on this work, the authors developed a new method to calculate the shadow enlargement including their elliptical shape.

This new method is the most rigorous and accurate procedure to date. The *Eclipse Almanac* uses it in the lunar eclipse predictions presented here.

## Explanation of Lunar Eclipse Figures

There are 22 eclipses of the Moon during the period 2041 to 2050. A figure for each eclipse is included.

Each figure consists of two diagrams. The first one depicts the Moon's path through Earth's penumbral and umbral shadows (with Celestial North up). The second is a map showing the geographic visibility of each eclipse phase. All features in these diagrams are identified in the key on the next page.

The Moon's orbital motion with respect to the shadows is from west to east (right to left). Each phase of the eclipse is defined by the instant when the Moon's limb is externally or internally tangent to the penumbra or umbra. The six primary contacts of the Moon with the penumbral and umbral shadows are defined as follows.

| | |
|---|---|
| **P1** — Penumbral Eclipse Begins | (Instant of first exterior tangency of the Moon with the Penumbra) |
| **U1** — Partial Eclipse Begins | (Instant of first exterior tangency of the Moon with the Umbra) |
| **U2** — Total Eclipse Begins | (Instant of first interior tangency of the Moon with the Umbra) |
| **U3** — Total Eclipse Ends | (Instant of last interior tangency of the Moon with the Umbra) |
| **U4** — Partial Eclipse Ends | (Instant of last exterior tangency of the Moon with the Umbra) |
| **P4** — Penumbral Eclipse Ends | (Instant of last exterior tangency of the Moon with the Penumbra) |

Penumbral lunar eclipses have two primary contacts: P1 and P4, but neither is observable because the edge of the penumbra is indistinct and extremely faint.

In addition to the penumbral contacts, partial lunar eclipses have two more contacts as the Moon's limb enters and exits the umbral shadow: U1 and U4, respectively. These two contacts mark the instants when the partial phase of the eclipse begins and ends.

Total lunar eclipses undergo all six contacts. The two additional umbral contacts are the instants when the Moon's entire disk is first and last internally tangent to the umbra: U2 and U3, respectively. They mark the times when the total phase of the eclipse begins and ends.

The Moon passes closest to the shadow axis at the instant of greatest eclipse. This corresponds to the maximum phase of the eclipse and the Moon's position at this instant is also depicted in the path diagrams.

The equidistant cylindrical projection map shows the geographic region of visibility at each phase of the eclipse. This is accomplished using a series of curves showing where Moonrise and Moonset occur at each eclipse contact. The map is shaded to indicate eclipse visibility. The entire eclipse is visible from the zone with no shading. Conversely, none of the eclipse can be seen from the zone with the darkest shading.

At greatest eclipse, the Moon is deepest in Earth's shadow. The geographic location where the Moon appears in the zenith at greatest eclipse is shown by an asterisk.

Parameters relevant to the eclipse appear in on the right side of each figure. The instant of greatest eclipse is the time (Universal Time[8] or UT1) when the Moon passes closest to the shadow axis. The penumbral and umbral magnitudes are the fractions of the Moon's diameter immersed in each shadow. Gamma is the minimum distance of the Moon's

---

[7] Herald, D., and Sinnott, R. W., "Analysis of Lunar Crater Timings, 1842–2011," *J. Br. Astron. Assoc.*, **124**, 5 (2014)

[8] Universal Time (UT1) is the modern-day replacement for Greenwich Mean Time and is based on Earth's rotation with respect to distant quasars.

center from the axis of Earth's shadow at greatest eclipse. The Saros series of the eclipse, and the node of the Moon's orbit are given. Each contact time of the Moon's edge with the penumbral and umbral shadows is listed in Universal Time (UT1). Depending on the eclipse type, the duration of the penumbral, partial or total phases are given.

*2–7 A time sequence shows the partial phases and totality during the total lunar eclipse of 2000 July 16 from Maui, Hawaii (Copyright ©2000 by Fred Espenak).*

## Lunar Eclipse Figures

### Key to Lunar Eclipse Figures

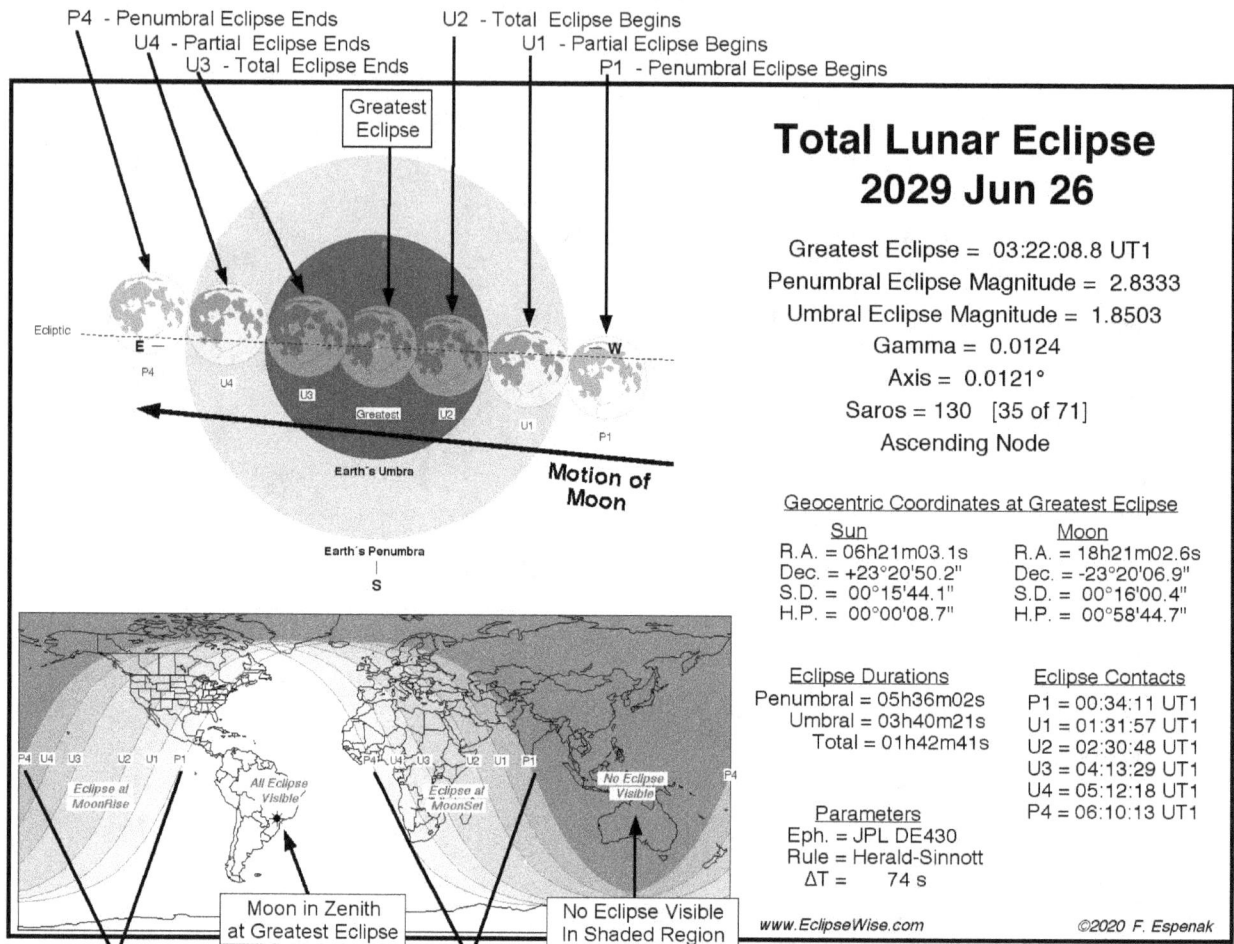

P4 - Penumbral Eclipse Ends
U4 - Partial Eclipse Ends
U3 - Total Eclipse Ends
U2 - Total Eclipse Begins
U1 - Partial Eclipse Begins
P1 - Penumbral Eclipse Begins

Greatest Eclipse

Ecliptic

E
P4
U4
U3
Greatest
U2
U1
W
P1

Earth's Umbra

Motion of Moon

Earth's Penumbra
S

**Total Lunar Eclipse
2029 Jun 26**

Greatest Eclipse = 03:22:08.8 UT1
Penumbral Eclipse Magnitude = 2.8333
Umbral Eclipse Magnitude = 1.8503
Gamma = 0.0124
Axis = 0.0121°
Saros = 130 [35 of 71]
Ascending Node

Geocentric Coordinates at Greatest Eclipse

| Sun | Moon |
|---|---|
| R.A. = 06h21m03.1s | R.A. = 18h21m02.6s |
| Dec. = +23°20'50.2" | Dec. = -23°20'06.9" |
| S.D. = 00°15'44.1" | S.D. = 00°16'00.4" |
| H.P. = 00°00'08.7" | H.P. = 00°58'44.7" |

| Eclipse Durations | Eclipse Contacts |
|---|---|
| Penumbral = 05h36m02s | P1 = 00:34:11 UT1 |
| Umbral = 03h40m21s | U1 = 01:31:57 UT1 |
| Total = 01h42m41s | U2 = 02:30:48 UT1 |
| | U3 = 04:13:29 UT1 |
| | U4 = 05:12:18 UT1 |
| Parameters | P4 = 06:10:13 UT1 |

Parameters
Eph. = JPL DE430
Rule = Herald-Sinnott
ΔT = 74 s

P4 U4 U3 U2 U1 P1
Eclipse at MoonRise
All Eclipse Visible
P4 U4 U3 U2 U1 P1
Eclipse at MoonSet
No Eclipse Visible
P4

Moon in Zenith at Greatest Eclipse

No Eclipse Visible In Shaded Region

www.EclipseWise.com                    ©2020 F. Espenak

**Eclipse During Moon Rise**
P1 - Penumbral Eclipse Begins
U1 - Partial Eclipse Begins
U2 - Total Eclipse Begins
U3 - Total Eclipse Ends
U4 - Partial Eclipse Ends
P4 - Penumbral Eclipse Ends

**Eclipse During Moon Set**
P1 - Penumbral Eclipse Begins
U1 - Partial Eclipse Begins
U2 - Total Eclipse Begins
U3 - Total Eclipse Ends
U4 - Partial Eclipse Ends
P4 - Penumbral Eclipse Ends

**Layout of Lunar Eclipse Figure**
**Upper Left:** Moon's Path Thru Earth's Shadows
**Lower Left:** Map of Eclipse Visibility
**Right Side:** Parameters for Lunar Eclipse

### Explanation of Parameters Used in Lunar Eclipse Figures

**Greatest Eclipse** – The instant when the Moon passes closest to the axis of Earth's shadow cone (Universal Time[9])
**Penumbral Eclipse Magnitude** – Fraction of the Moon's diameter immersed in the penumbra at greatest eclipse.
**Umbral Eclipse Magnitude** – The fraction of the Moon's diameter immersed in the umbra at greatest eclipse.
**Gamma** – Minimum distance from the Moon's center to Earth's shadow axis (units of Earth's equatorial radius).
**Axis** – Minimum distance from the Moon's center to Earth's shadow axis (units of degrees).
**Saros Series** – The Saros series that the eclipse belongs to. The numbers in "[ ]" are the eclipse's sequential position and the number of eclipses in the Saros series.
**Node** – The orbital node near which the eclipse takes place (Ascending Node or Descending Node).
**Geocentric Coordinates of the Sun and the Moon at Greatest Eclipse**
    **R.A.** – Right Ascension          **S.D.** – Semi-Diameter (i.e. - radius)
    **Dec.** – Declination              **H.P.** – Horizontal Parallax
**Eclipse Durations** – Durations of Penumbral, Partial, and Total Eclipse.
**Eclipse Contacts** – Contact Times (Universal Time or UT1) of the Moon with the Penumbra and the Umbra
    **P1, P4** – Start and End of the Penumbral Eclipse
    **U1, U4** – Start and End of the Partial Eclipse
    **U2, U3** – Start and End of the Total Eclipse

---

[9] Universal Time or UT1 is the modern replacement for Greenwich Mean Time

# Partial Lunar Eclipse
## 2041 May 16

Greatest Eclipse = 00:41:42.2 UT1
Penumbral Eclipse Magnitude = 1.0765
Umbral Eclipse Magnitude = 0.0663
Gamma = -0.9747
Axis = 0.9335°
Saros = 141 [25 of 72]
Decending Node

### Geocentric Coordinates at Greatest Eclipse

| Sun | Moon |
|---|---|
| R.A. = 03h32m49.6s | R.A. = 15h31m30.5s |
| Dec. = +19°08'35.5" | Dec. = -20°01'25.1" |
| S.D. = 00°15'49.2" | S.D. = 00°15'39.6" |
| H.P. = 00°00'08.7" | H.P. = 00°57'28.4" |

| Eclipse Durations | Eclipse Contacts |
|---|---|
| Penumbral = 04h30m30s | P1 = 22:26:22 UT1 |
| Umbral = 00h59m26s | U1 = 00:11:54 UT1 |
| | U4 = 01:11:20 UT1 |
| | P4 = 02:56:51 UT1 |

### Parameters
Eph. = JPL DE430
Rule = Herald-Sinnott
ΔT = 80 s

www.EclipseWise.com          ©2020 F. Espenak

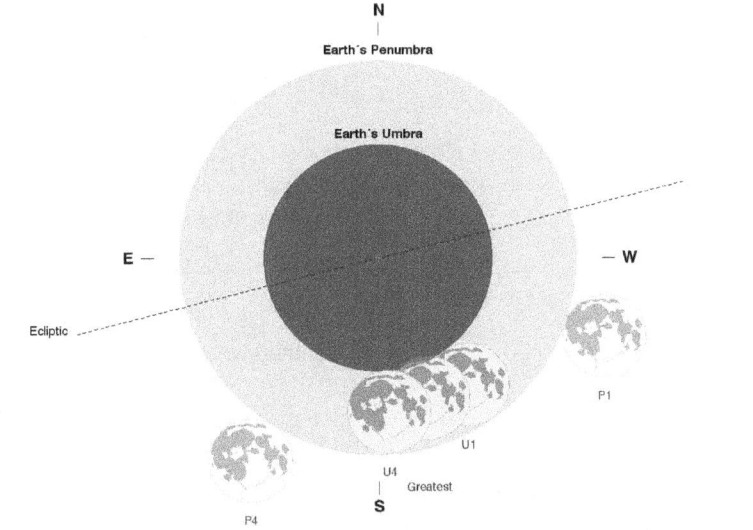

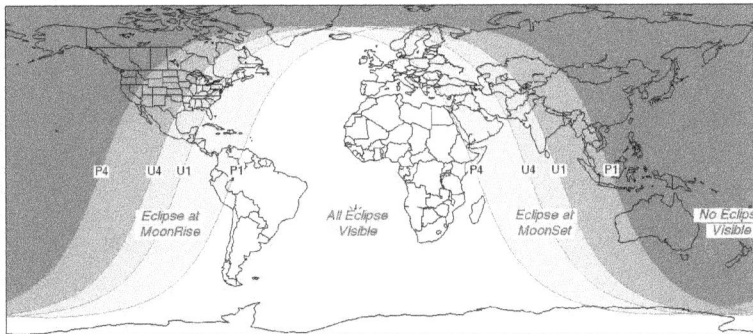

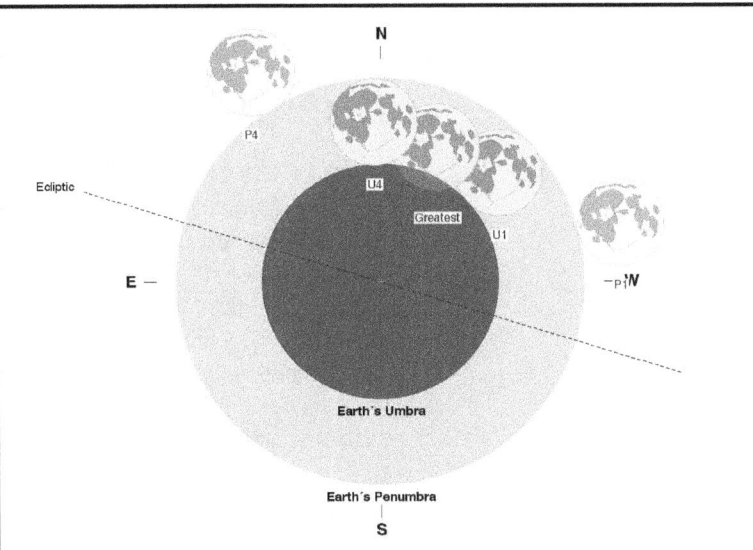

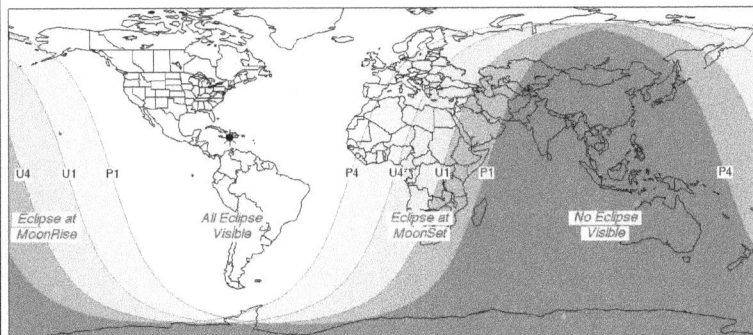

# Partial Lunar Eclipse
## 2041 Nov 08

Greatest Eclipse = 04:33:44.0 UT1
Penumbral Eclipse Magnitude = 1.1675
Umbral Eclipse Magnitude = 0.1714
Gamma = 0.9212
Axis = 0.9131°
Saros = 146 [12 of 72]
Ascending Node

### Geocentric Coordinates at Greatest Eclipse

| Sun | Moon |
|---|---|
| R.A. = 14h54m42.6s | R.A. = 02h53m15.3s |
| Dec. = -16°39'56.0" | Dec. = +17°30'36.2" |
| S.D. = 00°16'08.5" | S.D. = 00°16'12.4" |
| H.P. = 00°00'08.9" | H.P. = 00°59'28.8" |

| Eclipse Durations | Eclipse Contacts |
|---|---|
| Penumbral = 04h28m44s | P1 = 02:19:12 UT1 |
| Umbral = 01h31m04s | U1 = 03:47:54 UT1 |
| | U4 = 05:18:59 UT1 |
| | P4 = 06:47:57 UT1 |

### Parameters
Eph. = JPL DE430
Rule = Herald-Sinnott
ΔT = 80 s

www.EclipseWise.com          ©2020 F. Espenak

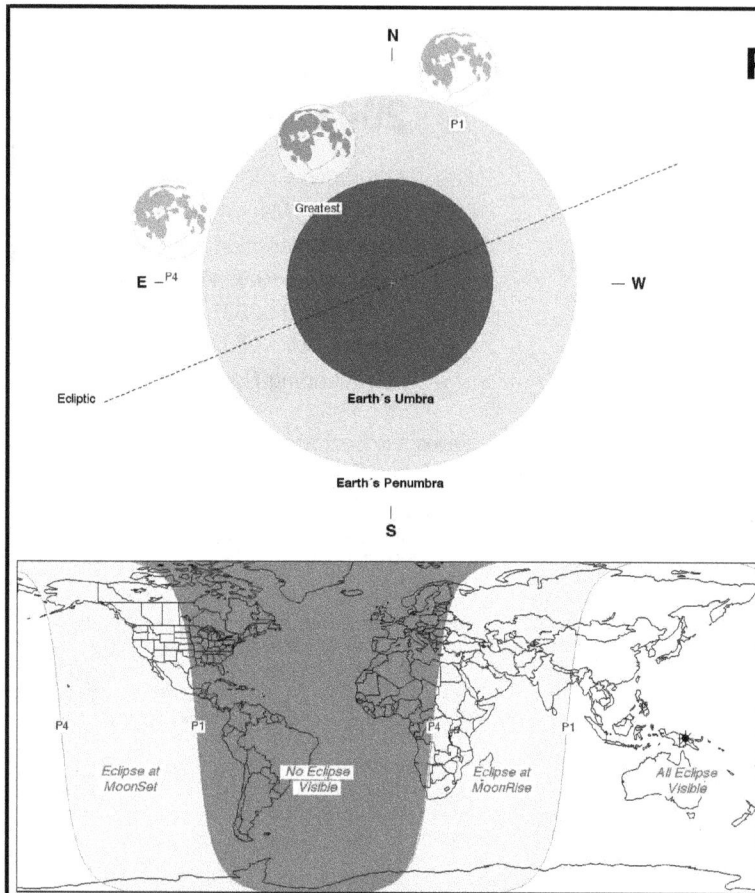

# Penumbral Lunar Eclipse
## 2042 Apr 05

Greatest Eclipse = 14:28:51.6 UT1
Penumbral Eclipse Magnitude = 0.8700
Umbral Eclipse Magnitude = -0.2156
Gamma = 1.1080
Axis = 0.9981°
Saros = 113 [65 of 71]
Decending Node

### Geocentric Coordinates at Greatest Eclipse

| Sun | Moon |
|---|---|
| R.A. = 00h58m43.2s | R.A. = 13h00m37.2s |
| Dec. = +06°16'08.8" | Dec. = -05°23'23.8" |
| S.D. = 00°15'59.3" | S.D. = 00°14'43.6" |
| H.P. = 00°00'08.8" | H.P. = 00°54'03.0" |

| Eclipse Durations | Eclipse Contacts |
|---|---|
| Penumbral = 04h29m11s | P1 = 12:14:30 UT1 |
| | P4 = 16:43:42 UT1 |

### Parameters
Eph. = JPL DE430
Rule = Herald-Sinnott
ΔT = 80 s

www.EclipseWise.com                    ©2020 F. Espenak

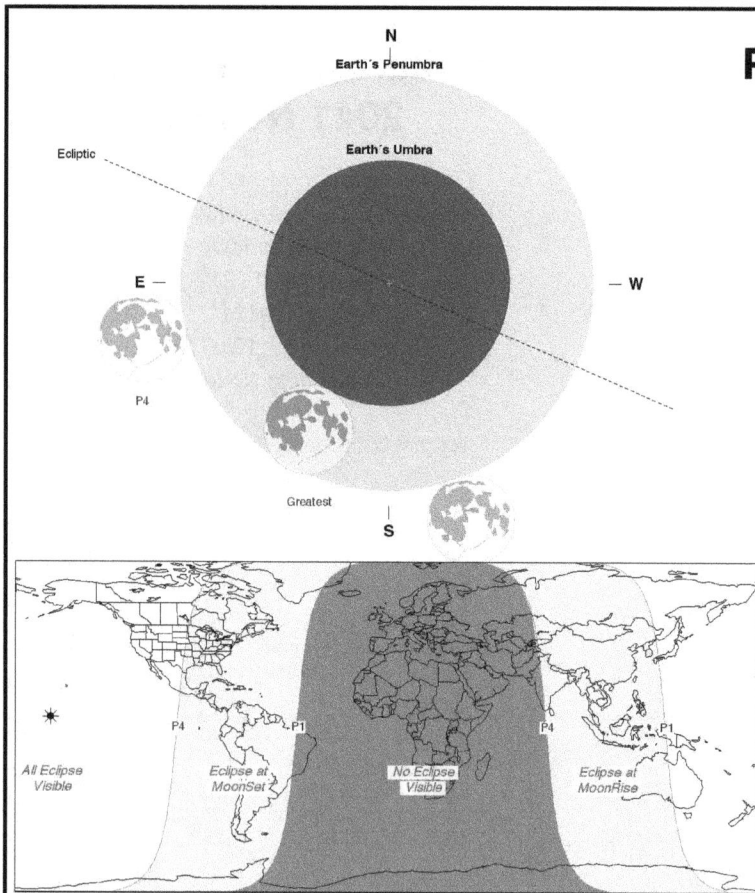

# Penumbral Lunar Eclipse
## 2042 Sep 29

Greatest Eclipse = 10:44:26.8 UT1
Penumbral Eclipse Magnitude = 0.9548
Umbral Eclipse Magnitude = -0.0010
Gamma = -1.0262
Axis = 1.0483°
Saros = 118 [53 of 73]
Ascending Node

### Geocentric Coordinates at Greatest Eclipse

| Sun | Moon |
|---|---|
| R.A. = 12h23m37.3s | R.A. = 00h25m38.7s |
| Dec. = -02°33'13.4" | Dec. = +01°38'07.3" |
| S.D. = 00°15'57.9" | S.D. = 00°16'42.1" |
| H.P. = 00°00'08.8" | H.P. = 01°01'18.0" |

| Eclipse Durations | Eclipse Contacts |
|---|---|
| Penumbral = 03h59m11s | P1 = 08:45:04 UT1 |
| | P4 = 12:44:15 UT1 |

### Parameters
Eph. = JPL DE430
Rule = Herald-Sinnott
ΔT = 81 s

www.EclipseWise.com                    ©2020 F. Espenak

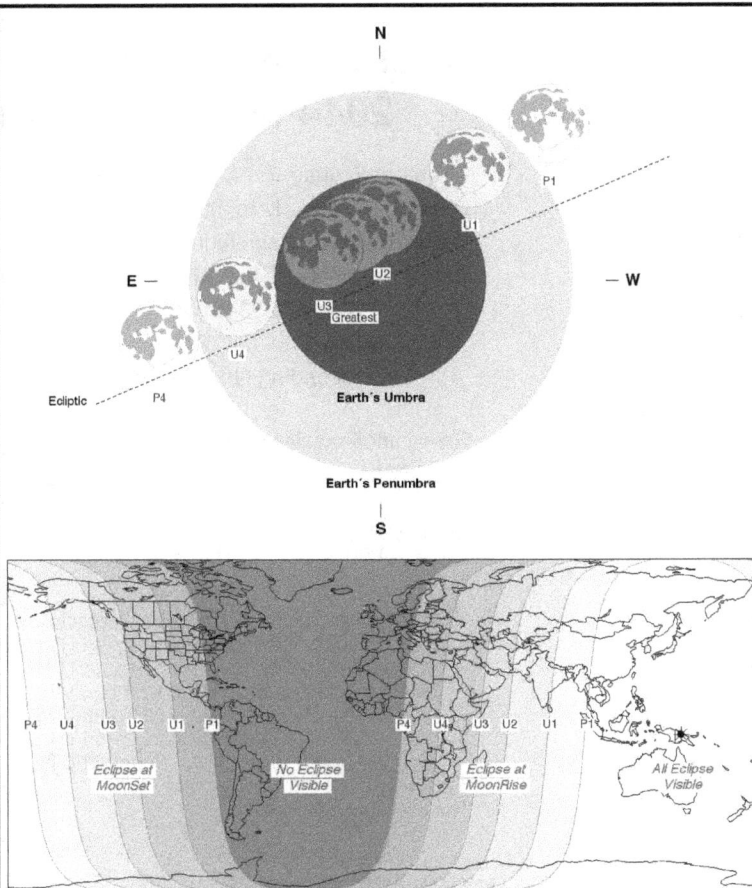

# Total Lunar Eclipse
# 2043 Mar 25

Greatest Eclipse = 14:30:42.6 UT1

Penumbral Eclipse Magnitude = 2.1920

Umbral Eclipse Magnitude = 1.1161

Gamma = 0.3849

Axis = 0.3510°

Saros = 123  [54 of 72]

Decending Node

### Geocentric Coordinates at Greatest Eclipse

| Sun | Moon |
|---|---|
| R.A. = 00h17m45.9s | R.A. = 12h18m26.9s |
| Dec. = +01°55'21.5" | Dec. = -01°36'57.6" |
| S.D. = 00°16'02.4" | S.D. = 00°14'54.5" |
| H.P. = 00°00'08.8" | H.P. = 00°54'42.9" |

| Eclipse Durations | Eclipse Contacts |
|---|---|
| Penumbral = 06h00m01s | P1 = 11:30:49 UT1 |
| Umbral = 03h35m18s | U1 = 12:43:10 UT1 |
| Total = 00h54m06s | U2 = 14:03:56 UT1 |
|  | U3 = 14:58:02 UT1 |
|  | U4 = 16:18:28 UT1 |
| Parameters | P4 = 17:30:50 UT1 |
| Eph. = JPL DE430 |  |
| Rule = Herald-Sinnott |  |
| ΔT =      81 s |  |

www.EclipseWise.com                    ©2020  F. Espenak

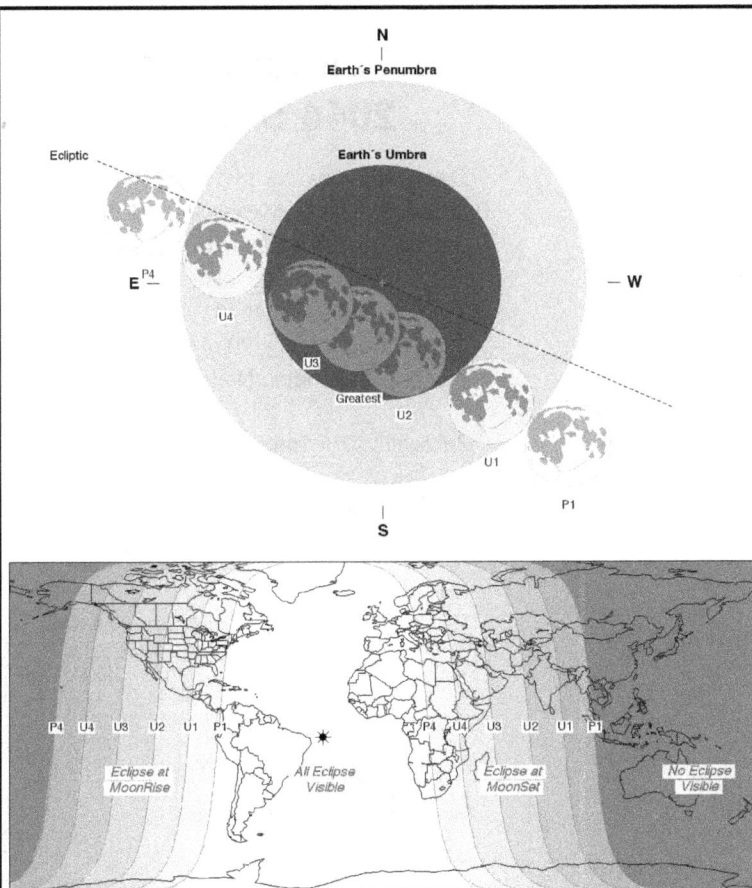

# Total Lunar Eclipse
# 2043 Sep 19

Greatest Eclipse = 01:50:28.6 UT1

Penumbral Eclipse Magnitude = 2.2452

Umbral Eclipse Magnitude = 1.2575

Gamma = -0.3316

Axis = 0.3269°

Saros = 128  [42 of 71]

Ascending Node

### Geocentric Coordinates at Greatest Eclipse

| Sun | Moon |
|---|---|
| R.A. = 11h45m28.0s | R.A. = 23h46m06.1s |
| Dec. = +01°34'24.4" | Dec. = -01°51'33.2" |
| S.D. = 00°15'55.1" | S.D. = 00°16'07.0" |
| H.P. = 00°00'08.8" | H.P. = 00°59'08.8" |

| Eclipse Durations | Eclipse Contacts |
|---|---|
| Penumbral = 05h26m27s | P1 = 23:07:17 UT1 |
| Umbral = 03h26m39s | U1 = 00:07:16 UT1 |
| Total = 01h12m20s | U2 = 01:14:33 UT1 |
|  | U3 = 02:26:52 UT1 |
|  | U4 = 03:33:56 UT1 |
| Parameters | P4 = 04:33:44 UT1 |
| Eph. = JPL DE430 |  |
| Rule = Herald-Sinnott |  |
| ΔT =      81 s |  |

www.EclipseWise.com                    ©2020  F. Espenak

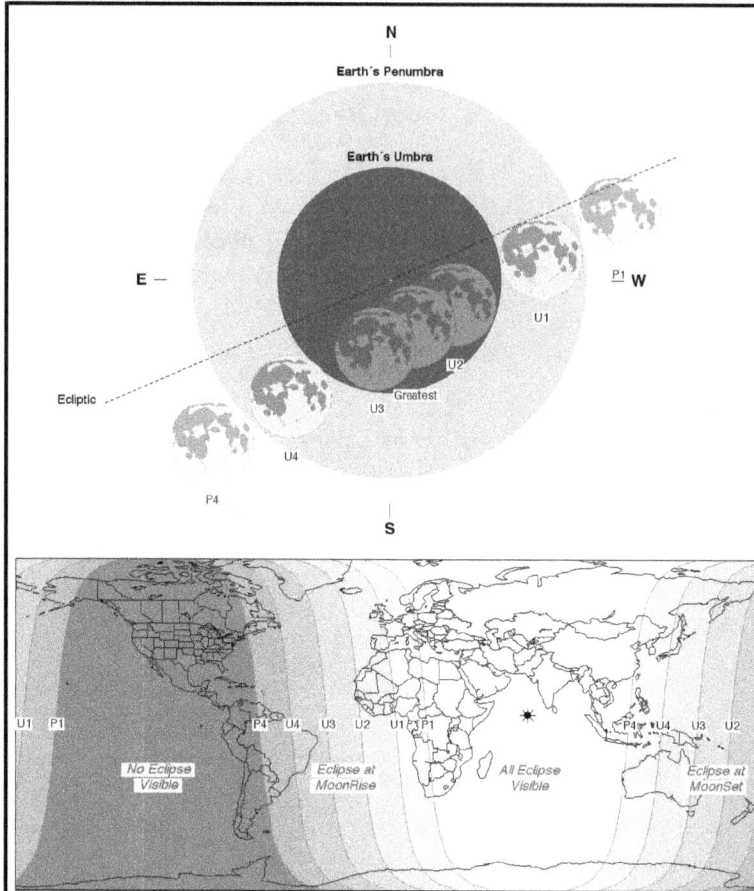

# Total Lunar Eclipse
## 2044 Mar 13

Greatest Eclipse = 19:37:11.0 UT1
Penumbral Eclipse Magnitude = 2.2322
Umbral Eclipse Magnitude = 1.2050
Gamma = -0.3496
Axis = 0.3349°
Saros = 133 [28 of 71]
Decending Node

### Geocentric Coordinates at Greatest Eclipse

| Sun | Moon |
|---|---|
| R.A. = 23h37m30.3s | R.A. = 11h36m51.3s |
| Dec. = -02°25'56.9" | Dec. = +02°08'22.5" |
| S.D. = 00°16'05.4" | S.D. = 00°15'39.8" |
| H.P. = 00°00'08.8" | H.P. = 00°57'29.1" |

| Eclipse Durations | Eclipse Contacts |
|---|---|
| Penumbral = 05h39m06s | P1 = 16:47:36 UT1 |
| Umbral = 03h29m43s | U1 = 17:52:11 UT1 |
| Total = 01h07m02s | U2 = 19:03:24 UT1 |
| | U3 = 20:10:26 UT1 |
| | U4 = 21:21:54 UT1 |
| Parameters | P4 = 22:26:42 UT1 |
| Eph. = JPL DE430 | |
| Rule = Herald-Sinnott | |
| ΔT = 82 s | |

www.EclipseWise.com    ©2020 F. Espenak

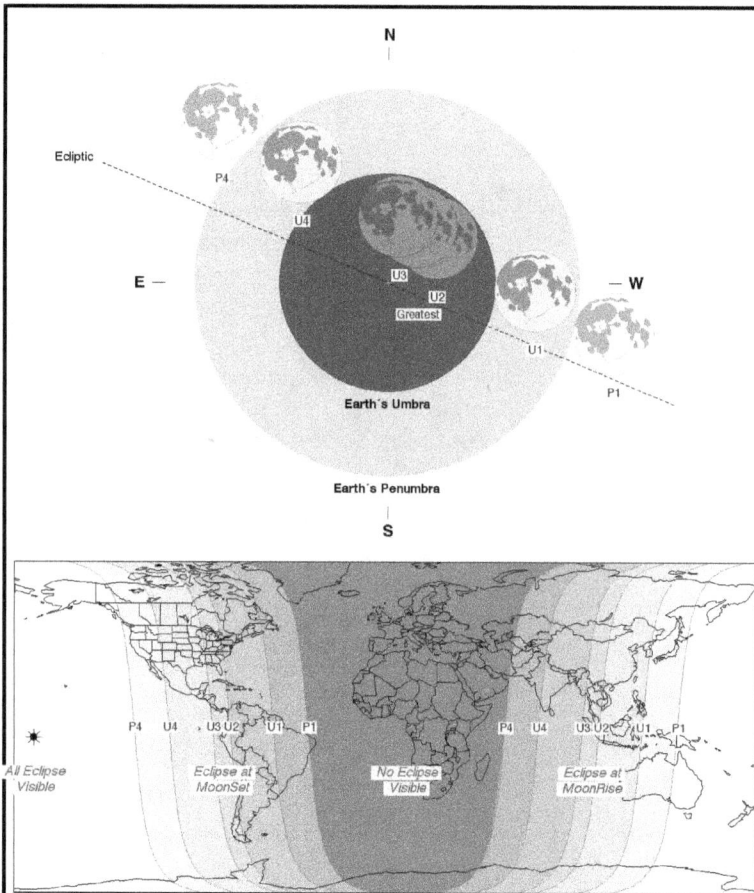

# Total Lunar Eclipse
## 2044 Sep 07

Greatest Eclipse = 11:19:22.6 UT1
Penumbral Eclipse Magnitude = 2.0879
Umbral Eclipse Magnitude = 1.0476
Gamma = 0.4318
Axis = 0.4030°
Saros = 138 [30 of 82]
Ascending Node

### Geocentric Coordinates at Greatest Eclipse

| Sun | Moon |
|---|---|
| R.A. = 11h06m33.5s | R.A. = 23h05m47.2s |
| Dec. = +05°43'12.4" | Dec. = -05°21'56.9" |
| S.D. = 00°15'52.4" | S.D. = 00°15'15.4" |
| H.P. = 00°00'08.7" | H.P. = 00°55'59.6" |

| Eclipse Durations | Eclipse Contacts |
|---|---|
| Penumbral = 05h44m46s | P1 = 08:26:52 UT1 |
| Umbral = 03h26m52s | U1 = 09:35:50 UT1 |
| Total = 00h34m47s | U2 = 11:01:42 UT1 |
| | U3 = 11:36:30 UT1 |
| | U4 = 13:02:42 UT1 |
| Parameters | P4 = 14:11:38 UT1 |
| Eph. = JPL DE430 | |
| Rule = Herald-Sinnott | |
| ΔT = 82 s | |

www.EclipseWise.com    ©2020 F. Espenak

# Penumbral Lunar Eclipse
## 2045 Mar 03

Greatest Eclipse = 07:42:02.8 UT1
Penumbral Eclipse Magnitude = 0.9643
Umbral Eclipse Magnitude = -0.0148
Gamma = -1.0274
Axis = 1.0354°
Saros = 143 [19 of 72]
Decending Node

### Geocentric Coordinates at Greatest Eclipse

| Sun | Moon |
|---|---|
| R.A. = 22h57m49.1s | R.A. = 10h55m51.5s |
| Dec. = -06°37'35.6" | Dec. = +05°42'46.0" |
| S.D. = 00°16'08.1" | S.D. = 00°16'28.7" |
| H.P. = 00°00'08.9" | H.P. = 01°00'28.6" |

| Eclipse Durations | Eclipse Contacts |
|---|---|
| Penumbral = 04h04m37s | P1 = 05:39:31 UT1 |
| | P4 = 09:44:08 UT1 |

### Parameters
Eph. = JPL DE430
Rule = Herald-Sinnott
ΔT = 82 s

www.EclipseWise.com                  ©2020 F. Espenak

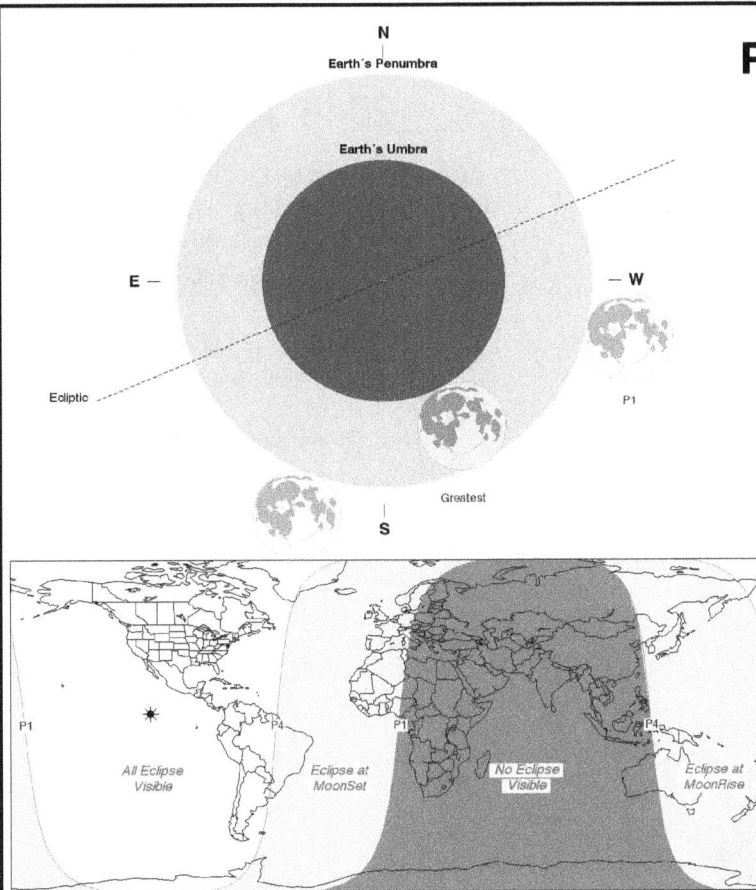

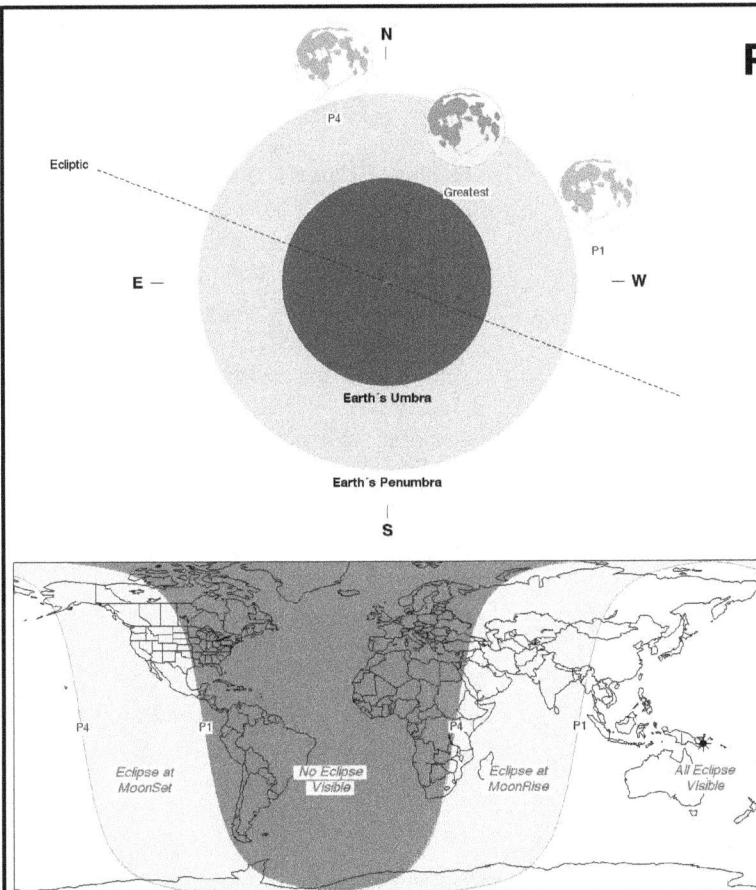

# Penumbral Lunar Eclipse
## 2045 Aug 27

Greatest Eclipse = 13:53:26.4 UT1
Penumbral Eclipse Magnitude = 0.6845
Umbral Eclipse Magnitude = -0.3899
Gamma = 1.2061
Axis = 1.0869°
Saros = 148 [5 of 70]
Ascending Node

### Geocentric Coordinates at Greatest Eclipse

| Sun | Moon |
|---|---|
| R.A. = 10h26m15.1s | R.A. = 22h24m15.1s |
| Dec. = +09°46'56.3" | Dec. = -08°48'49.2" |
| S.D. = 00°15'49.9" | S.D. = 00°14'44.1" |
| H.P. = 00°00'08.7" | H.P. = 00°54'04.7" |

| Eclipse Durations | Eclipse Contacts |
|---|---|
| Penumbral = 04h02m26s | P1 = 11:52:00 UT1 |
| | P4 = 15:54:26 UT1 |

### Parameters
Eph. = JPL DE430
Rule = Herald-Sinnott
ΔT = 82 s

www.EclipseWise.com                  ©2020 F. Espenak

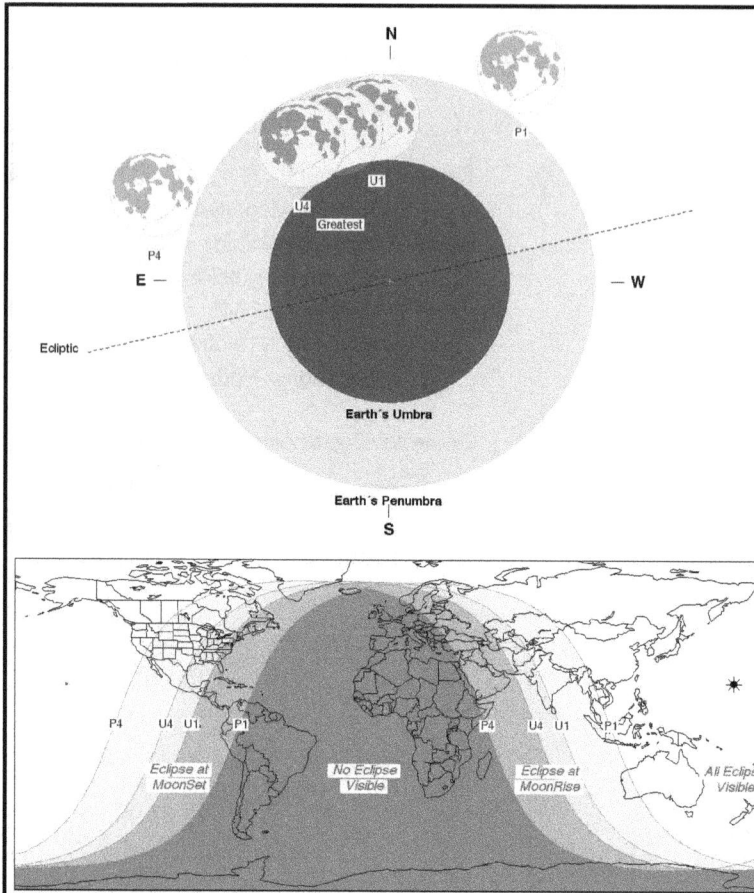

# Partial Lunar Eclipse
# 2046 Jan 22

Greatest Eclipse = 13:01:14.5 UT1
Penumbral Eclipse Magnitude = 1.0365
Umbral Eclipse Magnitude = 0.0550
Gamma = 0.9886
Axis = 1.0011°
Saros = 115 [59 of 72]
Decending Node

### Geocentric Coordinates at Greatest Eclipse

| Sun | Moon |
|---|---|
| R.A. = 20h19m45.5s | R.A. = 08h21m07.9s |
| Dec. = -19°33'42.8" | Dec. = +20°30'34.8" |
| S.D. = 00°16'15.1" | S.D. = 00°16'33.4" |
| H.P. = 00°00'08.9" | H.P. = 01°00'46.0" |

| Eclipse Durations | Eclipse Contacts |
|---|---|
| Penumbral = 04h10m45s | P1 = 10:56:01 UT1 |
| Umbral = 00h51m22s | U1 = 12:35:48 UT1 |
| | U4 = 13:27:10 UT1 |
| | P4 = 15:06:45 UT1 |

### Parameters
Eph. = JPL DE430
Rule = Herald-Sinnott
ΔT = 83 s

www.EclipseWise.com ©2020 F. Espenak

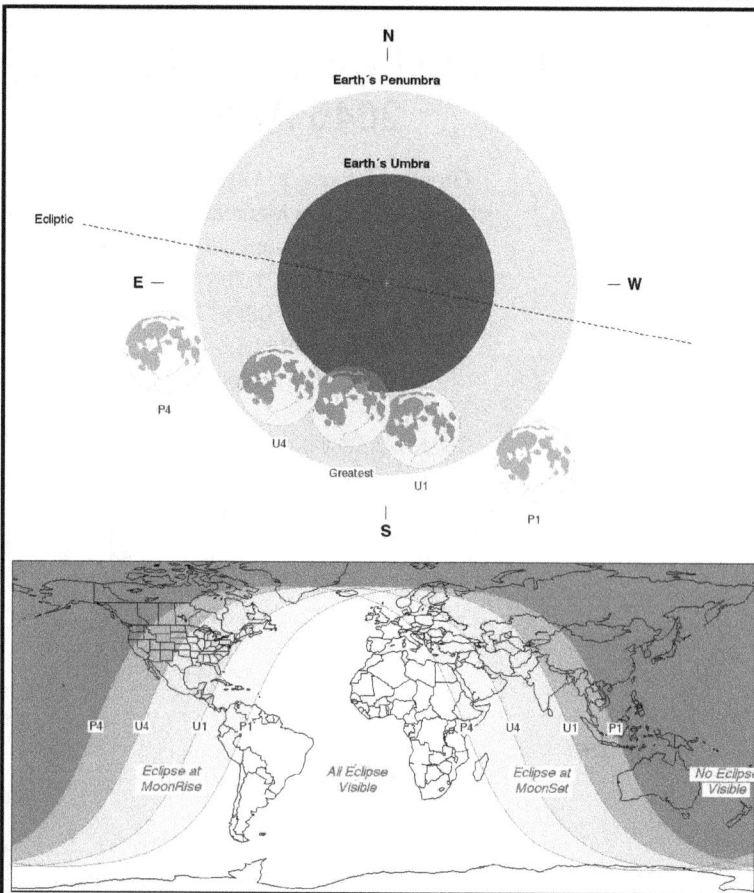

# Partial Lunar Eclipse
# 2046 Jul 18

Greatest Eclipse = 01:04:42.5 UT1
Penumbral Eclipse Magnitude = 1.2824
Umbral Eclipse Magnitude = 0.2478
Gamma = -0.8692
Axis = 0.8086°
Saros = 120 [59 of 83]
Ascending Node

### Geocentric Coordinates at Greatest Eclipse

| Sun | Moon |
|---|---|
| R.A. = 07h50m23.8s | R.A. = 19h51m22.3s |
| Dec. = +21°00'48.3" | Dec. = -21°47'22.3" |
| S.D. = 00°15'44.2" | S.D. = 00°15'12.7" |
| H.P. = 00°00'08.7" | H.P. = 00°55'49.5" |

| Eclipse Durations | Eclipse Contacts |
|---|---|
| Penumbral = 04h58m57s | P1 = 22:35:19 UT1 |
| Umbral = 01h55m18s | U1 = 00:07:08 UT1 |
| | U4 = 02:02:26 UT1 |
| | P4 = 03:34:16 UT1 |

### Parameters
Eph. = JPL DE430
Rule = Herald-Sinnott
ΔT = 83 s

www.EclipseWise.com ©2020 F. Espenak

# Total Lunar Eclipse
## 2047 Jan 12

Greatest Eclipse = 01:24:51.2 UT1
Penumbral Eclipse Magnitude = 2.2665
Umbral Eclipse Magnitude = 1.2358
Gamma = 0.3317
Axis = 0.3201°
Saros = 125 [50 of 72]
Decending Node

### Geocentric Coordinates at Greatest Eclipse

| Sun | Moon |
|---|---|
| R.A. = 19h33m56.9s | R.A. = 07h34m18.1s |
| Dec. = -21°40'46.3" | Dec. = +21°59'20.2" |
| S.D. = 00°16'15.8" | S.D. = 00°15'46.6" |
| H.P. = 00°00'08.9" | H.P. = 00°57'54.2" |

### Eclipse Durations
Penumbral = 05h38m05s
Umbral = 03h29m39s
Total = 01h10m41s

### Eclipse Contacts
P1 = 22:35:48 UT1
U1 = 23:40:07 UT1
U2 = 00:49:39 UT1
U3 = 02:00:20 UT1
U4 = 03:09:45 UT1
P4 = 04:13:53 UT1

### Parameters
Eph. = JPL DE430
Rule = Herald-Sinnott
ΔT = 83 s

www.EclipseWise.com    ©2020 F. Espenak

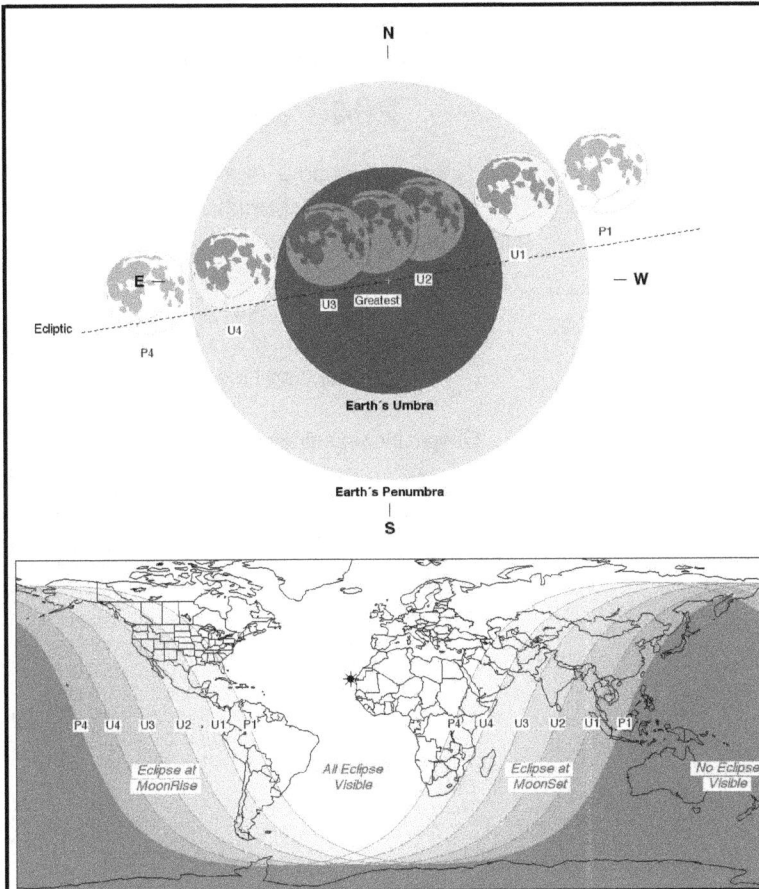

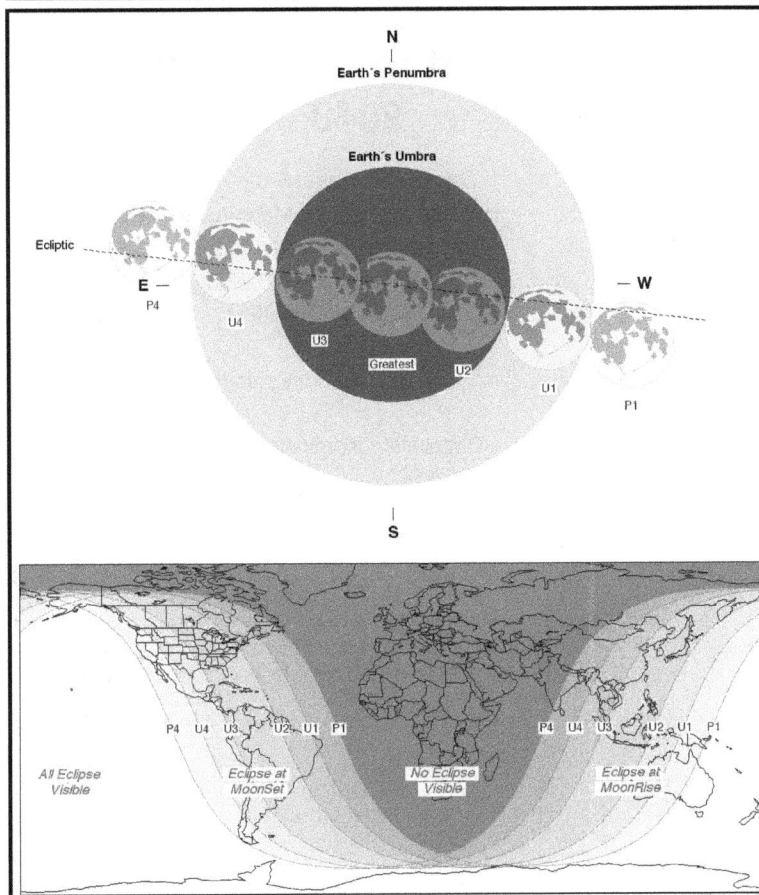

# Total Lunar Eclipse
## 2047 Jul 07

Greatest Eclipse = 10:34:21.9 UT1
Penumbral Eclipse Magnitude = 2.7326
Umbral Eclipse Magnitude = 1.7529
Gamma = -0.0636
Axis = 0.0625°
Saros = 130 [36 of 71]
Ascending Node

### Geocentric Coordinates at Greatest Eclipse

| Sun | Moon |
|---|---|
| R.A. = 07h06m19.6s | R.A. = 19h06m23.0s |
| Dec. = +22°33'30.9" | Dec. = -22°37'10.8" |
| S.D. = 00°15'43.9" | S.D. = 00°16'03.5" |
| H.P. = 00°00'08.7" | H.P. = 00°58'56.1" |

### Eclipse Durations
Penumbral = 05h34m19s
Umbral = 03h39m18s
Total = 01h41m35s

### Eclipse Contacts
P1 = 07:47:16 UT1
U1 = 08:44:42 UT1
U2 = 09:43:35 UT1
U3 = 11:25:10 UT1
U4 = 12:24:01 UT1
P4 = 13:21:34 UT1

### Parameters
Eph. = JPL DE430
Rule = Herald-Sinnott
ΔT = 84 s

www.EclipseWise.com    ©2020 F. Espenak

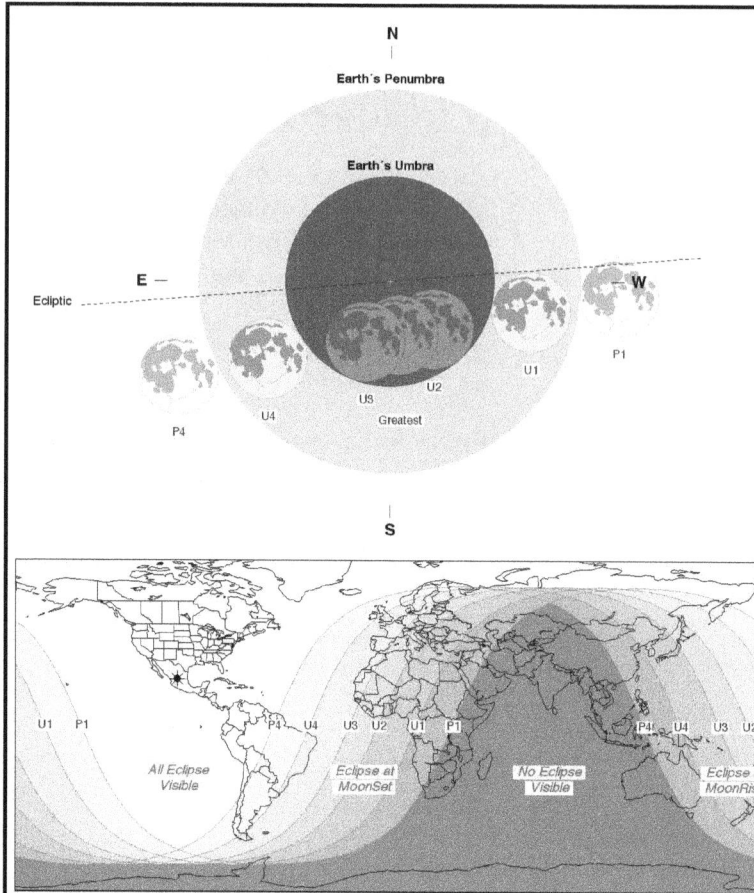

# Total Lunar Eclipse
# 2048 Jan 01

Greatest Eclipse = 06:52:30.9 UT1
Penumbral Eclipse Magnitude = 2.2158
Umbral Eclipse Magnitude = 1.1297
Gamma = -0.3746
Axis = 0.3431°
Saros = 135 [25 of 71]
Decending Node

### Geocentric Coordinates at Greatest Eclipse

| Sun | Moon |
|-----|------|
| R.A. = 18h45m45.0s | R.A. = 06h45m29.1s |
| Dec. = -23°01'00.1" | Dec. = +22°40'44.8" |
| S.D. = 00°16'15.9" | S.D. = 00°14'58.6" |
| H.P. = 00°00'08.9" | H.P. = 00°54'57.7" |

### Eclipse Durations
Penumbral = 06h00m21s
Umbral = 03h35m05s
Total = 00h56m40s

### Eclipse Contacts
P1 = 03:52:17 UT1
U1 = 05:04:57 UT1
U2 = 06:24:06 UT1
U3 = 07:20:46 UT1
U4 = 08:40:02 UT1
P4 = 09:52:38 UT1

### Parameters
Eph. = JPL DE430
Rule = Herald-Sinnott
ΔT = 84 s

www.EclipseWise.com                    ©2020  F. Espenak

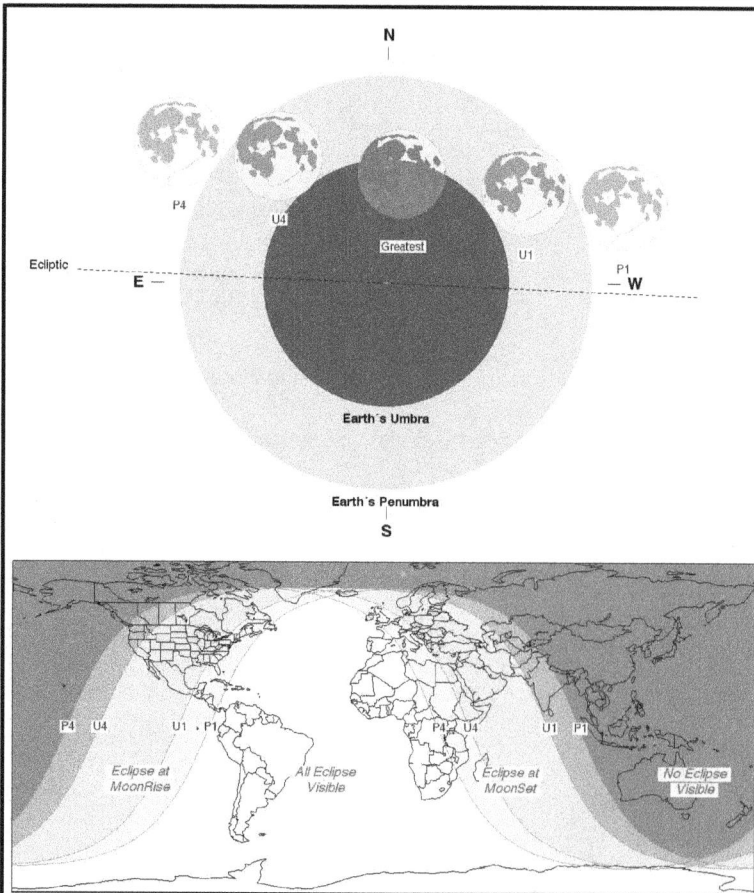

# Partial Lunar Eclipse
# 2048 Jun 26

Greatest Eclipse = 02:01:03.9 UT1
Penumbral Eclipse Magnitude = 1.5841
Umbral Eclipse Magnitude = 0.6404
Gamma = 0.6797
Axis = 0.6931°
Saros = 140 [26 of 77]
Ascending Node

### Geocentric Coordinates at Greatest Eclipse

| Sun | Moon |
|-----|------|
| R.A. = 06h22m31.9s | R.A. = 18h22m07.4s |
| Dec. = +23°19'54.0" | Dec. = -22°38'42.2" |
| S.D. = 00°15'44.1" | S.D. = 00°16'40.4" |
| H.P. = 00°00'08.7" | H.P. = 01°01'11.5" |

### Eclipse Durations
Penumbral = 04h46m32s
Umbral = 02h39m51s

### Eclipse Contacts
P1 = 23:37:46 UT1
U1 = 00:41:04 UT1
U4 = 03:20:55 UT1
P4 = 04:24:18 UT1

### Parameters
Eph. = JPL DE430
Rule = Herald-Sinnott
ΔT = 84 s

www.EclipseWise.com                    ©2020  F. Espenak

# Penumbral Lunar Eclipse
# 2048 Dec 20

Greatest Eclipse = 06:26:23.3 UT1
Penumbral Eclipse Magnitude = 0.9632
Umbral Eclipse Magnitude = -0.1420
Gamma = -1.0624
Axis = 0.9558°
Saros = 145 [13 of 71]
Decending Node

### Geocentric Coordinates at Greatest Eclipse

| Sun | Moon |
|---|---|
| R.A. = 17h55m49.3s | R.A. = 05h55m26.5s |
| Dec. = -23°25'43.8" | Dec. = +22°28'37.2" |
| S.D. = 00°16'15.4" | S.D. = 00°14'42.6" |
| H.P. = 00°00'08.9" | H.P. = 00°53'59.0" |

| Eclipse Durations | Eclipse Contacts |
|---|---|
| Penumbral = 04h42m26s | P1 = 04:05:07 UT1 |
| | P4 = 08:47:34 UT1 |

### Parameters
Eph. = JPL DE430
Rule = Herald-Sinnott
ΔT = 84 s

www.EclipseWise.com                    ©2020  F. Espenak

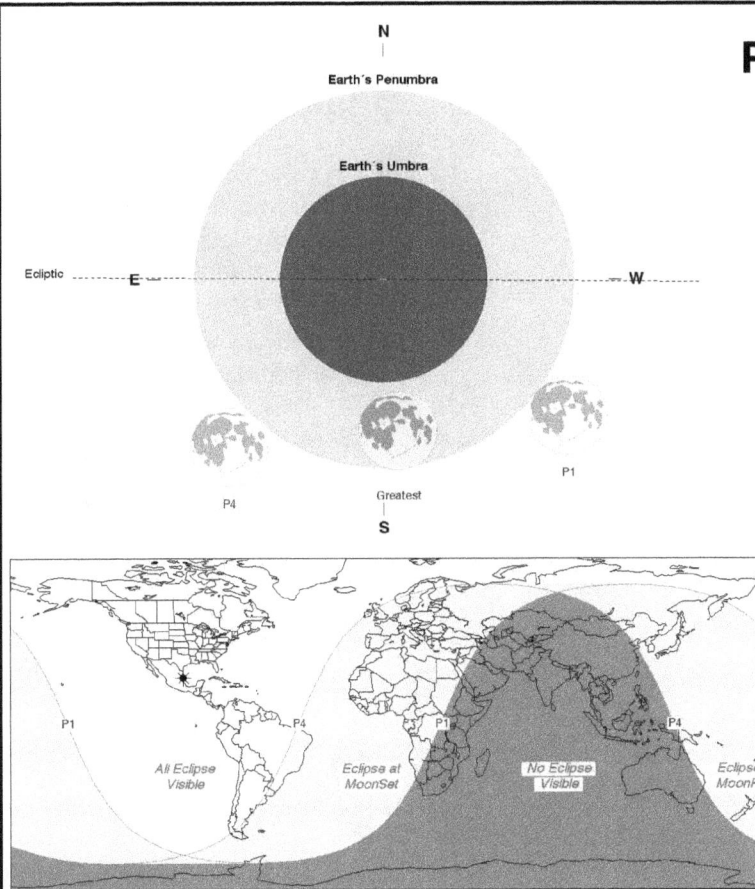

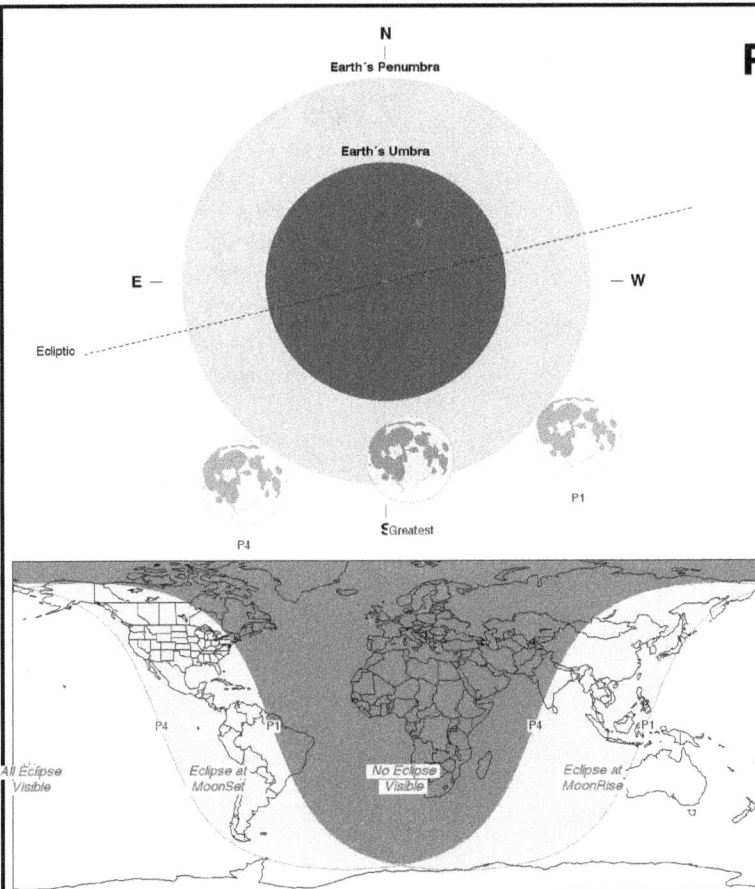

# Penumbral Lunar Eclipse
# 2049 May 17

Greatest Eclipse = 11:25:13.7 UT1
Penumbral Eclipse Magnitude = 0.7650
Umbral Eclipse Magnitude = -0.2073
Gamma = -1.1337
Axis = 1.1279°
Saros = 112 [67 of 72]
Ascending Node

### Geocentric Coordinates at Greatest Eclipse

| Sun | Moon |
|---|---|
| R.A. = 03h38m51.9s | R.A. = 15h38m12.8s |
| Dec. = +19°28'58.4" | Dec. = -20°36'01.8" |
| S.D. = 00°15'49.0" | S.D. = 00°16'16.0" |
| H.P. = 00°00'08.7" | H.P. = 00°59'41.9" |

| Eclipse Durations | Eclipse Contacts |
|---|---|
| Penumbral = 03h44m59s | P1 = 09:32:44 UT1 |
| | P4 = 13:17:43 UT1 |

### Parameters
Eph. = JPL DE430
Rule = Herald-Sinnott
ΔT = 85 s

www.EclipseWise.com                    ©2020  F. Espenak

# Penumbral Lunar Eclipse
# 2049 Jun 15

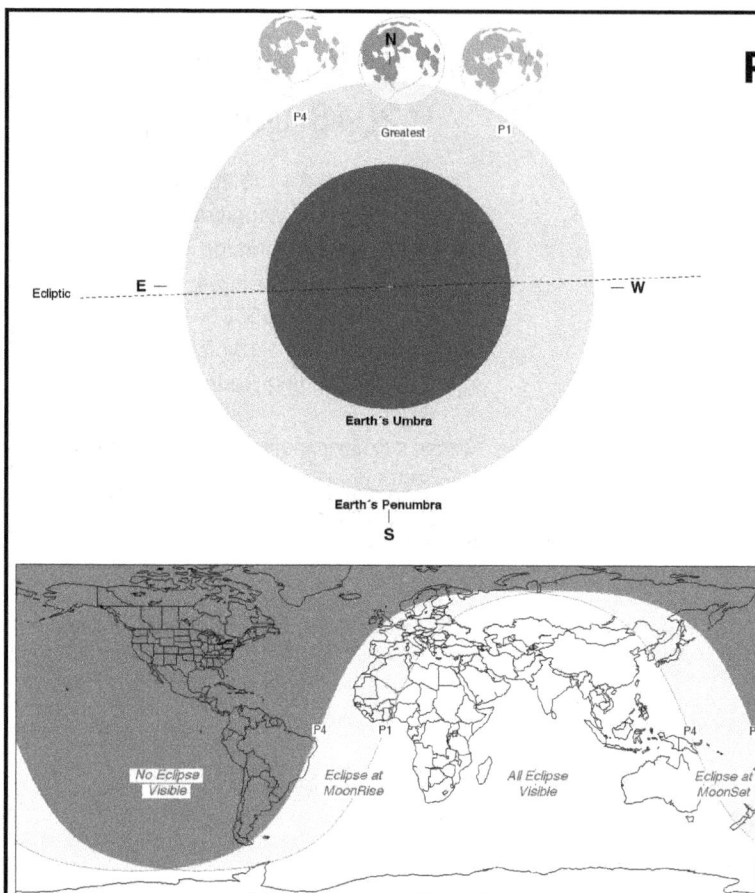

Greatest Eclipse = 19:12:46.8 UT1
Penumbral Eclipse Magnitude = 0.2526
Umbral Eclipse Magnitude = -0.6970
Gamma = 1.4069
Axis = 1.4268°
Saros = 150  [ 3 of 71]
Ascending Node

### Geocentric Coordinates at Greatest Eclipse

| Sun | Moon |
|---|---|
| R.A. = 05h38m45.5s | R.A. = 17h38m24.2s |
| Dec. = +23°20'31.0" | Dec. = -21°55'02.3" |
| S.D. = 00°15'44.8" | S.D. = 00°16'34.9" |
| H.P. = 00°00'08.7" | H.P. = 01°00'51.4" |

| Eclipse Durations | Eclipse Contacts |
|---|---|
| Penumbral = 02h12m41s | P1 = 18:06:29 UT1 |
| | P4 = 20:19:09 UT1 |

### Parameters
Eph. = JPL DE430
Rule = Herald-Sinnott
ΔT =    85 s

www.EclipseWise.com          ©2020  F. Espenak

---

# Penumbral Lunar Eclipse
# 2049 Nov 09

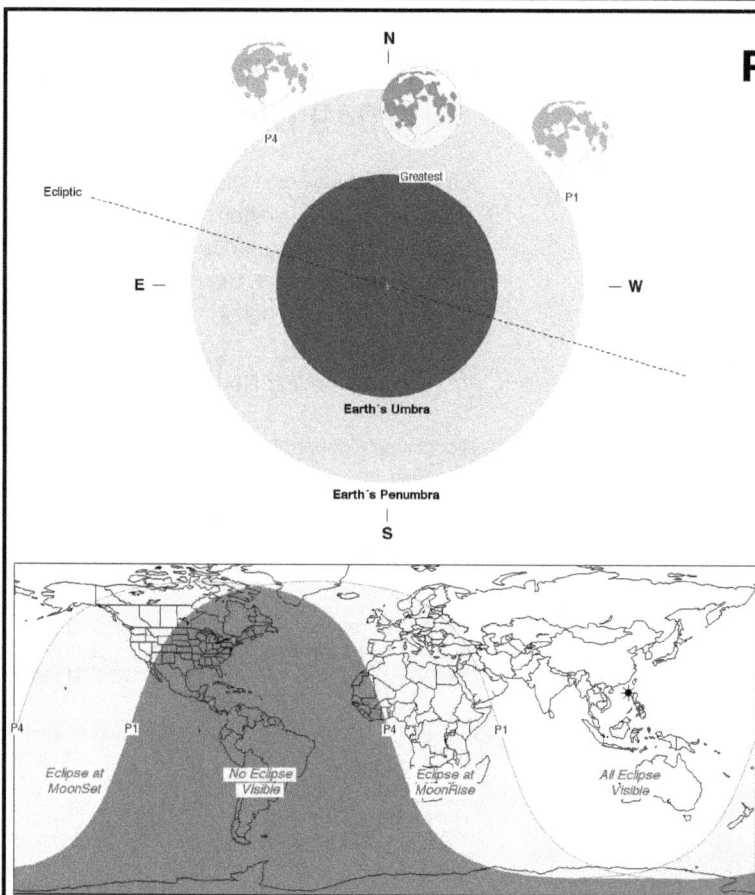

Greatest Eclipse = 15:50:45.7 UT1
Penumbral Eclipse Magnitude = 0.6821
Umbral Eclipse Magnitude = -0.3541
Gamma = 1.1965
Axis = 1.1404°
Saros = 117  [54 of 71]
Decending Node

### Geocentric Coordinates at Greatest Eclipse

| Sun | Moon |
|---|---|
| R.A. = 15h00m53.5s | R.A. = 03h00m00.0s |
| Dec. = -17°06'00.6" | Dec. = +18°13'14.6" |
| S.D. = 00°16'08.8" | S.D. = 00°15'35.1" |
| H.P. = 00°00'08.9" | H.P. = 00°57'11.8" |

| Eclipse Durations | Eclipse Contacts |
|---|---|
| Penumbral = 03h46m47s | P1 = 13:57:12 UT1 |
| | P4 = 17:43:59 UT1 |

### Parameters
Eph. = JPL DE430
Rule = Herald-Sinnott
ΔT =    85 s

www.EclipseWise.com          ©2020  F. Espenak

# Total Lunar Eclipse
# 2050 May 06

Greatest Eclipse = 22:30:36.3 UT1
Penumbral Eclipse Magnitude = 2.1064
Umbral Eclipse Magnitude = 1.0779
Gamma = -0.4181
Axis = 0.3942°
Saros = 122 [58 of 74]
Ascending Node

### Geocentric Coordinates at Greatest Eclipse

| Sun | Moon |
|---|---|
| R.A. = 02h56m30.8s | R.A. = 14h56m12.1s |
| Dec. = +16°47'28.5" | Dec. = -17°10'41.9" |
| S.D. = 00°15'51.3" | S.D. = 00°15'24.9" |
| H.P. = 00°00'08.7" | H.P. = 00°56'34.4" |

### Eclipse Durations
Penumbral = 05h40m55s
Umbral = 03h26m46s
Total = 00h43m49s

### Eclipse Contacts
P1 = 19:40:04 UT1
U1 = 20:47:12 UT1
U2 = 22:08:36 UT1
U3 = 22:52:25 UT1
U4 = 00:13:58 UT1
P4 = 01:20:59 UT1

### Parameters
Eph. = JPL DE430
Rule = Herald-Sinnott
ΔT = 85 s

www.EclipseWise.com                    ©2020 F. Espenak

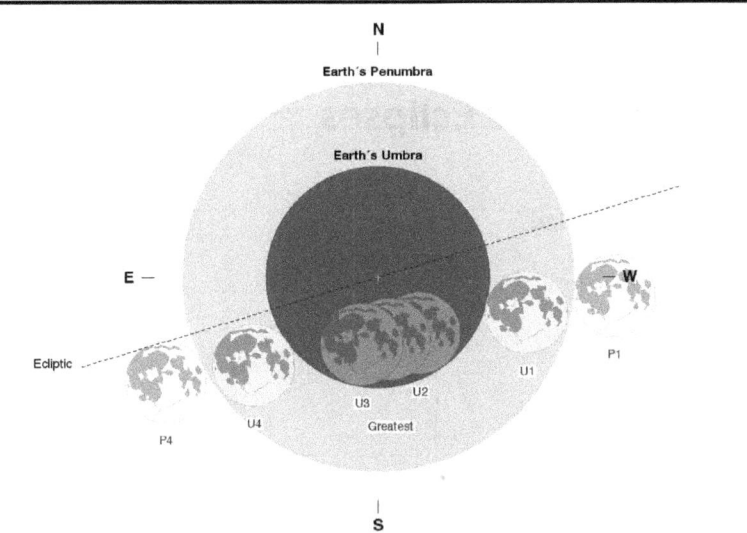

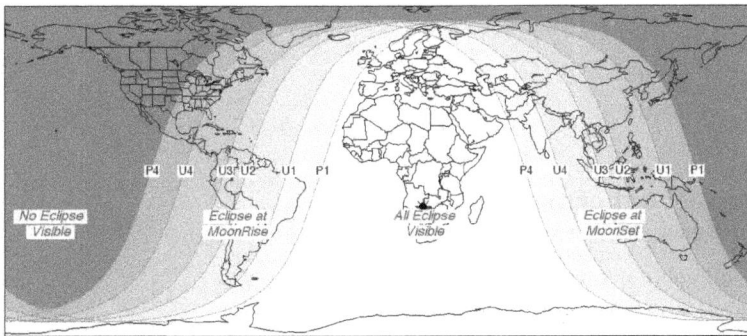

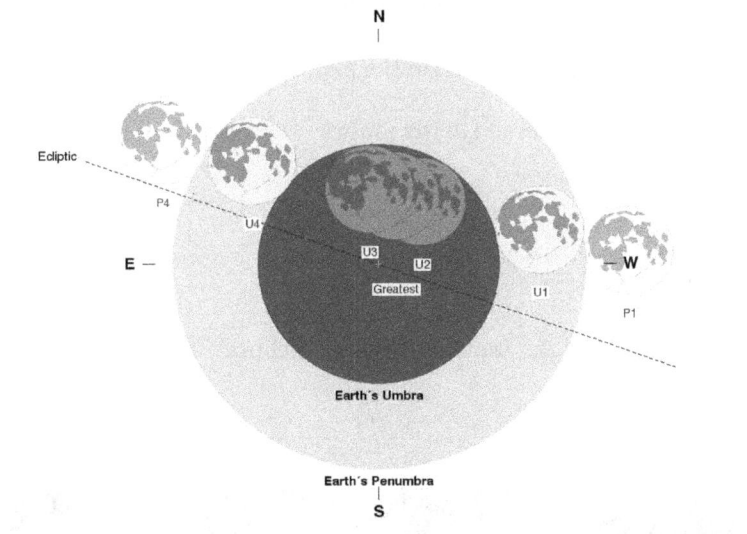

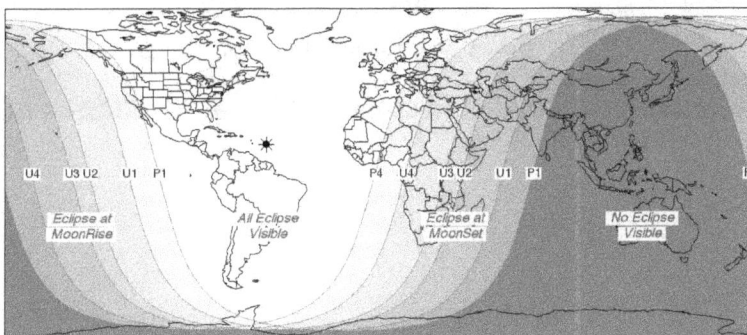

# Total Lunar Eclipse
# 2050 Oct 30

Greatest Eclipse = 03:20:21.2 UT1
Penumbral Eclipse Magnitude = 2.0356
Umbral Eclipse Magnitude = 1.0549
Gamma = 0.4435
Axis = 0.4454°
Saros = 127 [44 of 72]
Decending Node

### Geocentric Coordinates at Greatest Eclipse

| Sun | Moon |
|---|---|
| R.A. = 14h18m15.4s | R.A. = 02h17m49.7s |
| Dec. = -13°48'46.9" | Dec. = +14°14'46.2" |
| S.D. = 00°16'06.2" | S.D. = 00°16'25.2" |
| H.P. = 00°00'08.9" | H.P. = 01°00'15.6" |

### Eclipse Durations
Penumbral = 05h13m57s
Umbral = 03h13m34s
Total = 00h35m05s

### Eclipse Contacts
P1 = 00:43:22 UT1
U1 = 01:43:29 UT1
U2 = 03:02:39 UT1
U3 = 03:37:44 UT1
U4 = 04:57:03 UT1
P4 = 05:57:18 UT1

### Parameters
Eph. = JPL DE430
Rule = Herald-Sinnott
ΔT = 86 s

www.EclipseWise.com                    ©2020 F. Espenak

# Key to Catalog of Lunar Eclipses

The following catalog lists all lunar eclipses during a 25-year period.

A brief description of each parameter in the catalog appears below.

**Date** — Gregorian date of Greatest Eclipse

**Greatest Eclipse** — Universal Time of Greatest Eclipse

**Saros** — Saros Series Number of the eclipse

**Type** — Lunar Eclipse Type

> N = Penumbral Lunar Eclipse
> P = Partial Lunar Eclipse
> T = Total Lunar Eclipse
>
> + = Central total eclipse (Moon's center passes north of shadow axis)
> − = Central total eclipse (Moon's center passes south of shadow axis)
>
> b = Saros series begins (first penumbral eclipse in a Saros series)
> e = Saros series ends (last penumbral eclipse in a Saros series)

**Gamma** — minimum distance from center of the Moon to axis of Earth's umbral shadow

**P. Mag** — Penumbral Magnitude; fraction of the Moon's diameter immersed in the penumbra

**U. Mag** — Umbral Magnitude; fraction of the Moon's diameter immersed in the umbra

**Phase Durations**
> **Pen** — elapsed time from contact P1 to P4 (minutes)
> **Partial** — elapsed time from contact U1 to U4 (minutes)
> **Total** — elapsed time from contact U2 to U3 (minutes)

**Lat & Long** — latitude and longitude where the Moon appears in the zenith at Greatest Eclipse

*Photo 2–8 Time sequence of the total lunar eclipse of 2014 April 15. ©2014 F. Espenak*

# Catalog of Lunar Eclipses: 2041 to 2065

| Date | Greatest Eclipse | Saros | Type | Gamma | P. Mag | U.Mag | Pen | Partial | Total | Lat. | Long. |
|------|------------------|-------|------|-------|--------|-------|-----|---------|-------|------|-------|
| 2041 May 16 | 0:41:42 | 141 | P | -0.9747 | 1.0765 | 0.0663 | 270.5 | 59.4 | - | 20.0S | 11.7W |
| 2041 Nov 08 | 4:33:44 | 146 | P | 0.9212 | 1.1675 | 0.1714 | 268.7 | 91.1 | - | 17.5N | 72.9W |
| 2042 Apr 05 | 14:28:52 | 113 | N | 1.1080 | 0.8700 | -0.2156 | 269.2 | - | - | 5.4S | 143.9E |
| 2042 Sep 29 | 10:44:27 | 118 | N | -1.0262 | 0.9548 | -0.0010 | 239.2 | - | - | 1.6N | 163.0W |
| 2043 Mar 25 | 14:30:43 | 123 | T | 0.3849 | 2.1920 | 1.1161 | 360.0 | 215.3 | 54.1 | 1.6S | 144.0E |
| 2043 Sep 19 | 1:50:29 | 128 | T | -0.3316 | 2.2452 | 1.2575 | 326.4 | 206.7 | 72.3 | 1.9S | 29.0W |
| 2044 Mar 13 | 19:37:11 | 133 | T | -0.3496 | 2.2322 | 1.2050 | 339.1 | 209.7 | 67.0 | 2.1N | 67.8E |
| 2044 Sep 07 | 11:19:23 | 138 | T | 0.4318 | 2.0879 | 1.0476 | 344.8 | 206.9 | 34.8 | 5.4S | 170.6W |
| 2045 Mar 03 | 7:42:03 | 143 | N | -1.0274 | 0.9643 | -0.0148 | 244.6 | - | - | 5.7N | 113.0W |
| 2045 Aug 27 | 13:53:26 | 148 | N | 1.2061 | 0.6845 | -0.3899 | 242.4 | - | - | 8.8S | 151.5E |
| 2046 Jan 22 | 13:01:15 | 115 | P | 0.9886 | 1.0365 | 0.0550 | 250.7 | 51.4 | - | 20.5N | 167.9E |
| 2046 Jul 18 | 1:04:43 | 120 | P | -0.8692 | 1.2824 | 0.2478 | 299.0 | 115.3 | - | 21.8S | 14.4W |
| 2047 Jan 12 | 1:24:51 | 125 | T | 0.3317 | 2.2665 | 1.2358 | 338.1 | 209.6 | 70.7 | 22.0N | 19.1W |
| 2047 Jul 07 | 10:34:22 | 130 | Tm | -0.0636 | 2.7326 | 1.7529 | 334.3 | 219.3 | 101.6 | 22.6S | 157.3W |
| 2048 Jan 01 | 6:52:31 | 135 | T | -0.3746 | 2.2158 | 1.1297 | 360.4 | 215.1 | 56.7 | 22.7N | 102.4W |
| 2048 Jun 26 | 2:01:04 | 140 | P | 0.6796 | 1.5841 | 0.6404 | 286.5 | 159.8 | - | 22.6S | 29.6W |
| 2048 Dec 20 | 6:26:23 | 145 | N | -1.0624 | 0.9632 | -0.1420 | 282.4 | - | - | 22.5N | 97.3W |
| 2049 May 17 | 11:25:14 | 112 | N | -1.1337 | 0.7650 | -0.2073 | 225.0 | - | - | 20.6S | 172.4W |
| 2049 Jun 15 | 19:12:47 | 150 | N | 1.4069 | 0.2526 | -0.6970 | 132.7 | - | - | 21.9S | 71.9E |
| 2049 Nov 09 | 15:50:46 | 117 | N | 1.1965 | 0.6821 | -0.3541 | 226.8 | - | - | 18.2N | 118.1E |
| 2050 May 06 | 22:30:36 | 122 | T | -0.4181 | 2.1064 | 1.0779 | 340.9 | 206.8 | 43.8 | 17.2S | 21.4E |
| 2050 Oct 30 | 3:20:21 | 127 | T | 0.4435 | 2.0356 | 1.0549 | 313.9 | 193.6 | 35.1 | 14.2N | 54.3W |
| 2051 Apr 26 | 2:15:02 | 132 | T | 0.3371 | 2.2785 | 1.2034 | 365.7 | 221.7 | 70.3 | 13.2S | 34.2W |
| 2051 Oct 19 | 19:10:23 | 137 | T- | -0.2542 | 2.3719 | 1.4130 | 315.0 | 205.0 | 84.2 | 9.9N | 68.7E |
| 2052 Apr 14 | 2:16:40 | 142 | N | 1.0629 | 0.9478 | -0.1294 | 276.8 | - | - | 8.7S | 33.8W |
| 2052 Oct 08 | 10:44:31 | 147 | P | -0.9727 | 1.0653 | 0.0832 | 257.3 | 63.9 | - | 5.3N | 164.0W |
| 2053 Mar 04 | 17:20:42 | 114 | N | -1.0531 | 0.9334 | -0.0796 | 251.8 | - | - | 5.1N | 102.4E |
| 2053 Aug 29 | 8:04:22 | 119 | Nx | 1.0165 | 1.0203 | -0.0319 | 278.5 | - | - | 8.2S | 121.1W |
| 2054 Feb 22 | 6:49:59 | 124 | T | -0.3242 | 2.2502 | 1.2780 | 315.5 | 201.6 | 72.7 | 9.8N | 99.2W |
| 2054 Aug 18 | 9:25:02 | 129 | T | 0.2806 | 2.3817 | 1.3074 | 370.4 | 227.4 | 83.7 | 12.7S | 140.4W |
| 2055 Feb 11 | 22:44:49 | 134 | T | 0.3526 | 2.1982 | 1.2258 | 313.7 | 199.1 | 66.6 | 14.1N | 22.4W |
| 2055 Aug 07 | 10:51:49 | 139 | P | -0.4769 | 2.0081 | 0.9606 | 347.2 | 204.2 | - | 16.8S | 161.4W |
| 2056 Feb 01 | 12:24:37 | 144 | N | 1.0682 | 0.9069 | -0.1084 | 248.0 | - | - | 18.1N | 177.4E |
| 2056 Jun 27 | 10:01:40 | 111 | N | 1.3770 | 0.3158 | -0.6504 | 150.6 | - | - | 21.9S | 149.5W |
| 2056 Jul 26 | 18:41:55 | 149 | N | -1.2048 | 0.6448 | -0.3477 | 215.2 | - | - | 20.3S | 81.3E |
| 2056 Dec 22 | 1:47:26 | 116 | N | -1.1560 | 0.7872 | -0.3093 | 257.3 | - | - | 22.4N | 27.1W |
| 2057 Jun 17 | 2:24:50 | 121 | P | 0.6168 | 1.6982 | 0.7570 | 291.4 | 170.0 | - | 22.8S | 35.9W |
| 2057 Dec 11 | 0:52:08 | 126 | P | -0.4853 | 2.0194 | 0.9197 | 359.7 | 204.8 | - | 22.6N | 14.7W |
| 2058 Jun 06 | 19:14:17 | 131 | T- | -0.1181 | 2.6226 | 1.6628 | 324.5 | 214.1 | 98.0 | 22.8S | 71.1E |
| 2058 Nov 30 | 3:14:47 | 136 | T+ | 0.2208 | 2.4819 | 1.4277 | 353.9 | 221.5 | 90.4 | 21.9N | 51.6W |
| 2059 May 27 | 7:54:03 | 141 | P | -0.9098 | 1.1963 | 0.1846 | 282.5 | 97.9 | - | 22.2S | 119.5W |
| 2059 Nov 19 | 13:00:04 | 146 | P | 0.9004 | 1.2055 | 0.2097 | 271.2 | 99.9 | - | 20.4N | 161.0W |
| 2060 Apr 15 | 21:35:33 | 113 | N | 1.1622 | 0.7694 | -0.3136 | 255.7 | - | - | 9.3S | 36.5E |
| 2060 Oct 09 | 18:52:00 | 118 | N | -1.0671 | 0.8816 | -0.0779 | 232.0 | - | - | 5.8N | 74.3E |
| 2060 Nov 08 | 4:02:40 | 156 | Nb | 1.5332 | 0.0286 | -0.9356 | 45.3 | - | - | 18.2N | 65.4W |
| 2061 Apr 04 | 21:52:32 | 123 | T | 0.4300 | 2.1064 | 1.0360 | 355.8 | 210.3 | 30.9 | 5.8S | 32.7E |
| 2061 Sep 29 | 9:36:40 | 128 | T | -0.3810 | 2.1576 | 1.1640 | 325.5 | 203.1 | 59.6 | 2.4N | 146.4W |
| 2062 Mar 25 | 3:32:17 | 133 | T | -0.3150 | 2.2925 | 1.2715 | 339.1 | 212.0 | 75.3 | 2.2S | 51.7W |
| 2062 Sep 18 | 18:32:28 | 138 | T | 0.3736 | 2.1979 | 1.1515 | 350.0 | 213.1 | 60.2 | 1.2S | 80.2E |
| 2063 Mar 14 | 16:04:14 | 143 | P | -1.0008 | 1.0108 | 0.0363 | 248.5 | 41.9 | - | 1.5N | 120.7E |
| 2063 Sep 07 | 20:39:36 | 148 | N | 1.1375 | 0.8121 | -0.2657 | 261.2 | - | - | 4.9S | 49.1W |
| 2064 Feb 02 | 21:47:22 | 115 | P | 0.9969 | 1.0215 | 0.0395 | 249.8 | 43.7 | - | 17.6N | 37.0E |
| 2064 Jul 28 | 7:51:13 | 120 | P | -0.9473 | 1.1378 | 0.1055 | 285.0 | 76.5 | - | 19.6S | 115.8W |
| 2065 Jan 22 | 9:57:23 | 125 | T | 0.3371 | 2.2579 | 1.2248 | 339.0 | 209.7 | 69.4 | 19.8N | 146.3W |
| 2065 Jul 17 | 17:47:04 | 130 | T- | -0.1402 | 2.5907 | 1.6138 | 331.9 | 217.0 | 97.7 | 21.1S | 94.9E |

## Section 3: Phases of the Moon: 2041 to 2050

*Photo 3–1 Various phases of the Moon over one lunar month. ©2010 F. Espenak*

### Phases of the Moon

| Year | New Moon | First Quarter | Full Moon | Last Quarter |
|---|---|---|---|---|
| 2041 | Jan 02 19:08 | Jan 09 10:06 | Jan 17 07:11 | Jan 25 10:33 |
| | Feb 01 05:43 | Feb 07 23:40 | Feb 16 02:21 | Feb 24 00:29 |
| | Mar 02 15:39 | Mar 09 15:51 | Mar 17 20:19 | Mar 25 10:32 |
| | Apr 01 01:29 | Apr 08 09:38 | Apr 16 12:00 | Apr 23 17:24 |
| | Apr 30 11:46 T | May 08 03:54 | May 16 00:52 p | May 22 22:26 |
| | May 29 22:56 | Jun 06 21:40 | Jun 14 10:59 | Jun 21 03:12 |
| | Jun 28 11:17 | Jul 06 14:12 | Jul 13 19:01 | Jul 20 09:13 |
| | Jul 28 01:02 | Aug 05 04:53 | Aug 12 02:04 | Aug 18 17:43 |
| | Aug 26 16:16 | Sep 03 17:19 | Sep 10 09:24 | Sep 17 05:33 |
| | Sep 25 08:41 | Oct 03 03:33 | Oct 09 18:03 | Oct 16 21:05 |
| | Oct 25 01:30 A | Nov 01 12:05 | Nov 08 04:43 p | Nov 15 16:06 |
| | Nov 23 17:36 | Nov 30 19:49 | Dec 07 17:42 | Dec 15 13:32 |
| | Dec 23 08:06 | Dec 30 03:46 | | |

| Year | New Moon | First Quarter | Full Moon | Last Quarter |
|---|---|---|---|---|
| 2042 | | | Jan 06 08:54 | Jan 14 11:24 |
| | Jan 21 20:42 | Jan 28 12:48 | Feb 05 01:58 | Feb 13 07:16 |
| | Feb 20 07:39 | Feb 26 23:29 | Mar 06 20:10 | Mar 14 23:21 |
| | Mar 21 17:23 | Mar 28 12:00 | Apr 05 14:16 n | Apr 13 11:09 |
| | Apr 20 02:19 T | Apr 27 02:19 | May 05 06:48 | May 12 19:18 |
| | May 19 10:55 | May 26 18:18 | Jun 03 20:48 | Jun 11 01:00 |
| | Jun 17 19:48 | Jun 25 11:29 | Jul 03 08:09 | Jul 10 05:38 |
| | Jul 17 05:52 | Jul 25 05:01 | Aug 01 17:33 | Aug 08 10:35 |
| | Aug 15 18:01 | Aug 23 21:55 | Aug 31 02:02 | Sep 06 17:09 |
| | Sep 14 08:50 | Sep 22 13:20 | Sep 29 10:34 n | Oct 06 02:35 |
| | Oct 14 02:03 A | Oct 22 02:53 | Oct 28 19:48 | Nov 04 15:51 |
| | Nov 12 20:28 | Nov 20 14:31 | Nov 27 06:06 | Dec 04 09:19 |
| | Dec 12 14:29 | Dec 20 00:28 | Dec 26 17:43 | |

All times are in Universal Time (UT1).
Eclipses at New Moon (solar eclipse) or Full Moon (lunar eclipse), are indicated by these symbols:

| Solar Eclipses | Lunar Eclipses |
|---|---|
| T – Total | t – Total (Umbral) |
| A – Annular | p – Partial (Umbral) |
| H – Hybrid | n – Penumbral |
| P – Partial | |

# Phases of the Moon

| Year | New Moon | First Quarter | Full Moon | Last Quarter |
|------|----------|---------------|-----------|--------------|
| 2043 | | | | Jan 03  06:08 |
| | Jan 11  06:53 | Jan 18  09:05 | Jan 25  06:56 | Feb 02  04:14 |
| | Feb 09  21:07 | Feb 16  17:00 | Feb 23  21:58 | Mar 04  01:07 |
| | Mar 11  09:09 | Mar 18  01:03 | Mar 25  14:26 t | Apr 02  18:56 |
| | Apr 09  19:06 T | Apr 16  10:09 | Apr 24  07:23 | May 02  08:59 |
| | May 09  03:21 | May 15  21:05 | May 23  23:37 | May 31  19:25 |
| | Jun 07  10:35 | Jun 14  10:19 | Jun 22  14:20 | Jun 30  02:53 |
| | Jul 06  17:51 | Jul 14  01:47 | Jul 22  03:24 | Jul 29  08:23 |
| | Aug 05  02:23 | Aug 12  18:57 | Aug 20  15:04 | Aug 27  13:09 |
| | Sep 03  13:17 | Sep 11  13:01 | Sep 19  01:47 t | Sep 25  18:40 |
| | Oct 03  03:12 A | Oct 11  07:05 | Oct 18  11:56 | Oct 25  02:27 |
| | Nov 01  19:57 | Nov 10  00:13 | Nov 16  21:52 | Nov 23  13:46 |
| | Dec 01  14:37 | Dec 09  15:27 | Dec 16  08:02 | Dec 23  05:04 |
| | Dec 31  09:48 | | | |

| Year | New Moon | First Quarter | Full Moon | First Quarter |
|------|----------|---------------|-----------|--------------|
| 2044 | | Jan 08  04:02 | Jan 14  18:51 | Jan 21  23:47 |
| | Jan 30  04:04 | Feb 06  13:46 | Feb 13  06:42 | Feb 20  20:20 |
| | Feb 28  20:12 A | Mar 06  21:17 | Mar 13  19:41 t | Mar 21  16:52 |
| | Mar 29  09:26 | Apr 05  03:45 | Apr 12  09:39 | Apr 20  11:48 |
| | Apr 27  19:42 | May 04  10:28 | May 12  00:16 | May 20  04:02 |
| | May 27  03:39 | Jun 02  18:33 | Jun 10  15:16 | Jun 18  17:00 |
| | Jun 25  10:24 | Jul 02  04:48 | Jul 10  06:22 | Jul 18  02:47 |
| | Jul 24  17:10 | Jul 31  17:40 | Aug 08  21:14 | Aug 16  10:03 |
| | Aug 23  01:06 T | Aug 30  09:18 | Sep 07  11:24 t | Sep 14  15:58 |
| | Sep 21  11:03 | Sep 29  03:30 | Oct 07  00:30 | Oct 13  21:52 |
| | Oct 20  23:36 | Oct 28  23:28 | Nov 05  12:27 | Nov 12  05:09 |
| | Nov 19  14:58 | Nov 27  19:36 | Dec 04  23:34 | Dec 11  14:52 |
| | Dec 19  08:53 | Dec 27  14:00 | | |

| Year | New Moon | First Quarter | Full Moon | Last Quarter |
|------|----------|---------------|-----------|--------------|
| 2045 | | | Jan 03  10:20 | Jan 10  03:32 |
| | Jan 18  04:25 | Jan 26  05:09 | Feb 01  21:05 | Feb 08  19:03 |
| | Feb 16  23:51 A | Feb 24  16:37 | Mar 03  07:52 n | Mar 10  12:50 |
| | Mar 18  17:15 | Mar 26  00:56 | Apr 01  18:43 | Apr 09  07:52 |
| | Apr 17  07:27 | Apr 24  07:12 | May 01  05:52 | May 09  02:51 |
| | May 16  18:26 | May 23  12:38 | May 30  17:52 | Jun 07  20:23 |
| | Jun 15  03:05 | Jun 21  18:28 | Jun 29  07:16 | Jul 07  11:31 |
| | Jul 14  10:28 | Jul 21  01:52 | Jul 28  22:11 | Aug 05  23:57 |
| | Aug 12  17:39 T | Aug 19  11:55 | Aug 27  14:08 n | Sep 04  10:03 |
| | Sep 11  01:27 | Sep 18  01:30 | Sep 26  06:11 | Oct 03  18:31 |
| | Oct 10  10:37 | Oct 17  18:55 | Oct 25  21:31 | Nov 02  02:09 |
| | Nov 08  21:49 | Nov 16  15:26 | Nov 24  11:43 | Dec 01  09:46 |
| | Dec 08  11:41 | Dec 16  13:08 | Dec 24  00:49 | Dec 30  18:11 |

All times are in Universal Time (UT1).
Eclipses at New Moon (solar eclipse) or Full Moon (lunar eclipse), are indicated by these symbols:

| Solar Eclipses | Lunar Eclipses |
|----------------|----------------|
| T - Total | t - Total (Umbral) |
| A - Annular | p - Partial (Umbral) |
| H - Hybrid | n - Penumbral |
| P - Partial | |

# Phases of the Moon

| Year | New Moon | First Quarter | Full Moon | Last Quarter |
|------|----------|---------------|-----------|--------------|
| 2046 | Jan 07  04:24 | Jan 15  09:42 | Jan 22  12:51 p | Jan 29  04:11 |
|      | Feb 05  23:10 A | Feb 14  03:20 | Feb 20  23:44 | Feb 27  16:23 |
|      | Mar 07  18:15 | Mar 15  17:13 | Mar 22  09:27 | Mar 29  06:57 |
|      | Apr 06  11:52 | Apr 14  03:21 | Apr 20  18:21 | Apr 27  23:30 |
|      | May 06  02:56 | May 13  10:25 | May 20  03:15 | May 27  17:06 |
|      | Jun 04  15:22 | Jun 11  15:27 | Jun 18  13:10 | Jun 26  10:40 |
|      | Jul 04  01:39 | Jul 10  19:53 | Jul 18  00:55 p | Jul 26  03:19 |
|      | Aug 02  10:25 T | Aug 09  01:15 | Aug 16  14:50 | Aug 24  18:36 |
|      | Aug 31  18:25 | Sep 07  09:07 | Sep 15  06:39 | Sep 23  08:16 |
|      | Sep 30  02:25 | Oct 06  20:41 | Oct 14  23:41 | Oct 22  20:07 |
|      | Oct 29  11:17 | Nov 05  12:28 | Nov 13  17:04 | Nov 21  06:10 |
|      | Nov 27  21:50 | Dec 05  07:56 | Dec 13  09:55 | Dec 20  14:43 |
|      | Dec 27  10:39 | | | |

| Year | New Moon | First Quarter | Full Moon | Last Quarter |
|------|----------|---------------|-----------|--------------|
| 2047 | | Jan 04  05:31 | Jan 12  01:21 t | Jan 18  22:32 |
|      | Jan 26  01:44 P | Feb 03  03:09 | Feb 10  14:40 | Feb 17  06:42 |
|      | Feb 24  18:26 | Mar 04  22:52 | Mar 12  01:37 | Mar 18  16:11 |
|      | Mar 26  11:44 | Apr 03  15:11 | Apr 10  10:35 | Apr 17  03:30 |
|      | Apr 25  04:40 | May 03  03:26 | May 09  18:24 | May 16  16:46 |
|      | May 24  20:27 | Jun 01  11:54 | Jun 08  02:05 | Jun 15  07:45 |
|      | Jun 23  10:36 P | Jun 30  17:37 | Jul 07  10:33 t | Jul 15  00:09 |
|      | Jul 22  22:49 P | Jul 29  22:03 | Aug 05  20:38 | Aug 13  17:34 |
|      | Aug 21  09:16 | Aug 28  02:49 | Sep 04  08:54 | Sep 12  11:18 |
|      | Sep 19  18:31 | Sep 26  09:29 | Oct 03  23:42 | Oct 12  04:22 |
|      | Oct 19  03:28 | Oct 25  19:13 | Nov 02  16:58 | Nov 10  19:39 |
|      | Nov 17  12:59 | Nov 24  08:41 | Dec 02  11:55 | Dec 10  08:29 |
|      | Dec 16  23:38 P | Dec 24  01:51 | | |

| Year | New Moon | First Quarter | Full Moon | Last Quarter |
|------|----------|---------------|-----------|--------------|
| 2048 | | | Jan 01  06:57 t | Jan 08  18:49 |
|      | Jan 15  11:32 | Jan 22  21:56 | Jan 31  00:14 | Feb 07  03:16 |
|      | Feb 14  00:31 | Feb 21  19:22 | Feb 29  14:38 | Mar 07  10:45 |
|      | Mar 14  14:28 | Mar 22  16:03 | Mar 30  02:04 | Apr 05  18:10 |
|      | Apr 13  05:20 | Apr 21  10:02 | Apr 28  11:13 | May 05  02:22 |
|      | May 12  20:58 | May 21  00:16 | May 27  18:57 | Jun 03  12:05 |
|      | Jun 11  12:50 A | Jun 19  10:49 | Jun 26  02:08 p | Jul 02  23:58 |
|      | Jul 11  04:04 | Jul 18  18:31 | Jul 25  09:34 | Aug 01  14:30 |
|      | Aug 09  17:59 | Aug 17  00:32 | Aug 23  18:07 | Aug 31  07:42 |
|      | Sep 08  06:24 | Sep 15  06:04 | Sep 22  04:46 | Sep 30  02:45 |
|      | Oct 07  17:45 | Oct 14  12:20 | Oct 21  18:25 | Oct 29  22:15 |
|      | Nov 06  04:38 | Nov 12  20:29 | Nov 20  11:20 | Nov 28  16:34 |
|      | Dec 05  15:30 T | Dec 12  07:29 | Dec 20  06:39 n | Dec 28  08:31 |

All times are in Universal Time (UT1).
Eclipses at New Moon (solar eclipse) or Full Moon (lunar eclipse), are indicated by these symbols:

| Solar Eclipses | Lunar Eclipses |
|----------------|----------------|
| T – Total | t – Total (Umbral) |
| A – Annular | p – Partial (Umbral) |
| H – Hybrid | n – Penumbral |
| P – Partial | |

# Phases of the Moon

| Year | New Moon | First Quarter | Full Moon | Last Quarter |
|---|---|---|---|---|
| 2049 | Jan 04 02:24 | Jan 10 21:56 | Jan 19 02:29 | Jan 26 21:33 |
| | Feb 02 13:16 | Feb 09 15:38 | Feb 17 20:47 | Feb 25 07:36 |
| | Mar 04 00:11 | Mar 11 11:26 | Mar 19 12:23 | Mar 26 15:10 |
| | Apr 02 11:39 | Apr 10 07:27 | Apr 18 01:04 | Apr 24 21:11 |
| | May 02 00:11 | May 10 01:57 | May 17 11:13 n | May 24 02:54 |
| | May 31 14:00 A | Jun 08 17:56 | Jun 15 19:26 n | Jun 22 09:41 |
| | Jun 30 04:50 | Jul 08 07:10 | Jul 15 02:29 | Jul 21 18:48 |
| | Jul 29 20:07 | Aug 06 17:52 | Aug 13 09:19 | Aug 20 07:10 |
| | Aug 28 11:18 | Sep 05 02:28 | Sep 11 17:04 | Sep 18 23:03 |
| | Sep 27 02:05 | Oct 04 09:39 | Oct 11 02:53 | Oct 18 17:55 |
| | Oct 26 16:15 | Nov 02 16:19 | Nov 09 15:38 n | Nov 17 14:32 |
| | Nov 25 05:35 H | Dec 01 23:39 | Dec 09 07:28 | Dec 17 11:15 |
| | Dec 24 17:51 | Dec 31 08:53 | | |

| Year | New Moon | First Quarter | Full Moon | Last Quarter |
|---|---|---|---|---|
| 2050 | | | Jan 08 01:39 | Jan 16 06:17 |
| | Jan 23 04:57 | Jan 29 20:48 | Feb 06 20:47 | Feb 14 22:10 |
| | Feb 21 15:03 | Feb 28 11:29 | Mar 08 15:23 | Mar 16 10:08 |
| | Mar 23 00:41 | Mar 30 04:17 | Apr 07 08:12 | Apr 14 18:24 |
| | Apr 21 10:25 | Apr 28 22:08 | May 06 22:26 t | May 14 00:04 |
| | May 20 20:51 H | May 28 16:04 | Jun 05 09:51 | Jun 12 04:39 |
| | Jun 19 08:22 | Jun 27 09:17 | Jul 04 18:51 | Jul 11 09:46 |
| | Jul 18 21:17 | Jul 27 01:05 | Aug 03 02:20 | Aug 09 16:48 |
| | Aug 17 11:47 | Aug 25 14:56 | Sep 01 09:30 | Sep 08 02:51 |
| | Sep 16 03:49 | Sep 24 02:34 | Sep 30 17:31 | Oct 07 16:32 |
| | Oct 15 20:48 | Oct 23 12:10 | Oct 30 03:16 t | Nov 06 09:57 |
| | Nov 14 13:41 P | Nov 21 20:25 | Nov 28 15:09 | Dec 06 06:27 |
| | Dec 14 05:18 | Dec 21 04:15 | Dec 28 05:15 | |

All times are in Universal Time (UT1).
Eclipses at New Moon (solar eclipse) or Full Moon (lunar eclipse), are indicated by these symbols:

| Solar Eclipses | Lunar Eclipses |
|---|---|
| T – Total | t – Total (Umbral) |
| A – Annular | p – Partial (Umbral) |
| H – Hybrid | n – Penumbral |
| P – Partial | |

*Photo 3–2 The changing phases of the Moon. ©2010 F. Espenak*

The phases of the Moon for the entire 21st Century can be found at the AstroPixels.com website:

*www.astropixels.com/ephemeris/moon/phases2001gmt.html*

# Books by the Author

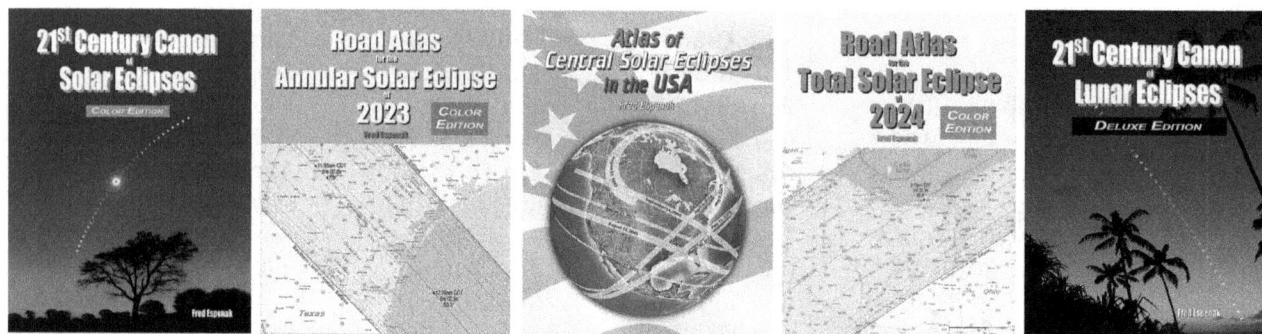

Below is a partial list of some of the books Fred Espenak has written through Astropixels Publishing:

### 21st Century Canon of Solar Eclipses

The complete guide to every solar eclipse occurring from 2001 tom 2100 (224 eclipses in all). It includes information and maps for all total, annular, hybrid, and partial eclipses. A special world atlas shows detailed full page maps of all central eclipse paths (total, annular and hybrid).

### Road Atlas for the Annular Solar Eclipse of 2023

Detailed road maps of the entire eclipse path from the western USA, through Mexico, Central and South America. Information printed on the maps makes it easy to estimate the duration of annularity from any location in the eclipse path. This is the next annular solar eclipse visible from the USA.

### Atlas of Central Solar Eclipses in the USA

When was the last total eclipse through the USA and when is the next? How often do they happen? What total eclipse tracks passed across the USA during the 17th, 18th, and 19th centuries, etc., and what states did they include? And how often is a total solar eclipse visible from each of the 50 states? The Atlas of Central Solar Eclipses in the USA answers all of these questions and more with hundreds of maps and tables.

### Road Atlas for the Total Solar Eclipse of 2024

This book contains detailed road maps of the entire eclipse path from Mexico, through the USA and Canada. Information printed on the maps makes it easy to estimate the duration of totality from any location in the eclipse path. This is the next total solar eclipse visible from the USA.

### 21st Century Canon of Lunar Eclipses

The complete guide to every lunar eclipse occurring from 2001 tom 2100 (228 eclipses in all). It includes information and maps for all total, partial, and penumbral eclipses. The predictions use a new model for Earth's elliptical shadows.

For information on these books and more, visit *Astropixels Publishing*:

*http://eclipsewise.com/pubs/index.html*